I0055753

Dragonflies Production Technology

The Author

Dr. Sathe Tukaram Vithalrao [M.Sc., Ph.D., Sangit Vishard, IBT (Seri.), F.I.S.E.C., F.S.E.Sc., F.S.L.Sc., F.I.C.C.B., F.S.S.I.] is presently working as Professor and Head, Department of Zoology, Shivaji University, Kolhapur. He has teaching experience of 29 years in Entomology at University PG department and 15 years in Agrochemicals and Pest Management. He has written 30 books and published 255 research papers in national and international journals of repute. He guided 20 Ph.D. students and completed 6 major research projects (from CSIR, DST, DBT and UGC). He visited Canada (1988), Japan (1988), Thailand (2002, 2004), Spain (2005), France (2005), South Korea (2006) and Nepal (2007) etc. for academic work. He is member of editorial board of eleven prestigious journals. He delivered 35 talks through All India Radio and internal conferences and involved in Doordarshan, S.T.V. and B. T.V. programmes on biodiversity of moths and butterflies. He published more than 35 popular articles in daily newspapers on insects and sericulture. He got several prestigious awards like "Environmentalists of the Year-2003", "Bharat Jyoti", "Jewel of India", "International Gold Star", "Eminent Citizen of India", "Education Acumen", "Best Educationist", "Eminent Scientist of the Year-2008", "Lifetime Education Achievement", "Lifetime Achievement in Entomology and Insect Taxonomy-2009", Educational Leadership-2011, Asia Pacific International Award-2012 etc. He is also working as Research and Recognition (RR) Committee member for Pune University, Pune; North Maharashtra University, Jalgaon; Shivaji University, Kolhapur and DBA Marathwada University, Aurangabad. He has been also awarded several fellowships from different scientific and academic societies. He is Chairman of Maharashtra District Environmental Centre of NESA.

Dragonflies Production Technology

T.V. Sathe
Head
Department of Zoology
Shivaji University
Kolhapur – 416 004, M.S.

2013
Daya Publishing House®
A Division of
Astral International Pvt. Ltd.
New Delhi – 110 002

© 2013 AUTHOR
ISBN 9789351300816

Publisher's note:

Every possible effort has been made to ensure that the information contained in this book is accurate at the time of going to press, and the publisher and author cannot accept responsibility for any errors or omissions, however caused. No responsibility for loss or damage occasioned to any person acting, or refraining from action, as a result of the material in this publication can be accepted by the editor, the publisher or the author. The Publisher is not associated with any product or vendor mentioned in the book. The contents of this work are intended to further general scientific research, understanding and discussion only. Readers should consult with a specialist where appropriate.

Every effort has been made to trace the owners of copyright material used in this book, if any. The author and the publisher will be grateful for any omission brought to their notice for acknowledgement in the future editions of the book.

All Rights reserved under International Copyright Conventions. No part of this publication may be reproduced, stored in a retrieval system, or transmitted in any form or by any means, electronic, mechanical, photocopying, recording or otherwise without the prior written consent of the publisher and the copyright owner.

Acknowledgment

It is a matter of great pleasure to express my sincere thanks to Secretary and Under Secretary, Ministry of Science and Technology, Govt. of India and DBT New Delhi authority for giving sanction to the present project.

I express my deep sense of gratitude toward Dr. Manjunathan, Task force committee member of DBT, New Delhi for his valuable suggestions.

My Sincere thanks are due to Shivaji University, Kolhapur authority for providing facilities for completion of this project.

I am specially thankful to Prof. V. V. Rammurty IARI, New Delhi and Prof. P. K. Nikam DBAMU, Aurangabad for their valuable discussions on the topic.

I am also thankful to Mr. A.R. Bhusnar, Project Assistant for his valuable assistance throughout the project work,

Lastly, I thanks all my well wishers

Prof. (Dr.) T.V. Sathe

Preface

Dragonflies are very good biocontrol agents of mosquitoes, jassids and several lepidopterous pests. Hence, production techniques for dragonflies, mosquitoes, paramecium and daphnia are given. Control of mosquitoes and other insect pests with conventional pesticides is difficult task. Secondly, pests develop resistance against various insecticides. Thirdly, pesticides lead serious problems like air and water pollution, health hazards, killing of beneficial organisms, pest resurgence, secondary pest outbreak and pest resistance. Hence, biological control is need of the day. Dragonflies can solve the problem of mosquito control and other pests by keeping environment eco-friendly.

For completion of this book and the results presented in the text, DBT, New Delhi, Ministry of Science and Technology, Govt. of India provided financial assistance to the project. Author is thankful to DBT and Ministry of Science and Technology, Govt. of India and also Shivaji University authority for providing facilities for this work.

I feel that the book will be stimulatory to the students, teachers, farmers, medicinal practitioners and scientists involved in control of insect pests by biological means.

Prof. (Dr.) T.V. Sathe

Contents

Chapter 1
Introduction

Dragonflies developing in aquatic ecosystem have tremendous importance in biological control of insect pests such as mosquitoes, jassids, midges and several kinds of moth in agro and forest ecosystems.

Odonates are commonly found darting and dancing actively near ponds, pools, rivers, streams and also marshy place (Prasad and Kulkarni, 2001). Some species of Odonata are also recorded perching high on trees and shrubs, considerable away from water and in dense forests. They are reported from sea levels to over 3, 600 m and from brackish marshy areas to desert lands from all over the world.

Out of 6, 500 species reported from the world, 550 species belonging to 139 genera of 17 families have been reported from India (Kulkarni and Prasad, 2002; Sathe and Shinde, 2008). Prasad and Kulkarni (2001) reported 71 species from Nilgiri Biosphere reserve and its environments. Prasad and Kulkarni (2002) also reported additional 34 species from Kerala. From Western Ghats 138 species have been reported. Most of the studies are confined to Western Ghats of Kerala and very little attention is paid on the Western Ghats scattered in Maharashtra except the recent work of Sathe and Shinde (2006, 2007,2008). Therefore, attempts have been made on biodiversity of dragonflies from Kolhapur district.

The dragonfly genera *Crocothemis, Pantala, Bradinopyga, Brachythemis, Tramea, Sympetrum* and *Czocontenemis* are associated with mosquito larvae for feeding and are effective biocontol agents of *Culex quinquefaciatus, Anopheles sinensis* and *Aedes aegypti.*

Mosquitoes cause dreaded diseases such as Malaria, Filaria, Dengue, Chikunguniya. Japanese encephalitis, etc. Therefore, they have tremendous importance in epidemiology. Mosquitoes are very difficult to control because they have developed resistance to pesticidal groups like chlorinated hydrocarbons, Organophosphorus compounds, carbamates and pyrethroids. Secondly, pesticides lead several other problems like health hazards, killing of beneficial organisms, secondary pest outbreak, pest resistance, pest resurgence, interruption to ecocycles, pollutions etc. This indicates that these is need to find out an alternative for chemical control of mosquitoes. Dragonflies are potential biocontrol agents of mosquitoes. Both mosquito larvae and dragonfly naiads develop in aquatic habitat. The naiads largely feed on mosquito larvae in aquatic habitat and control mosquito populations. In past, several workers (Metcalf and Flint, 1979; Thomas *et al.,* 1988; Santhamarina and Mijares 1986; Urable *et al.,*

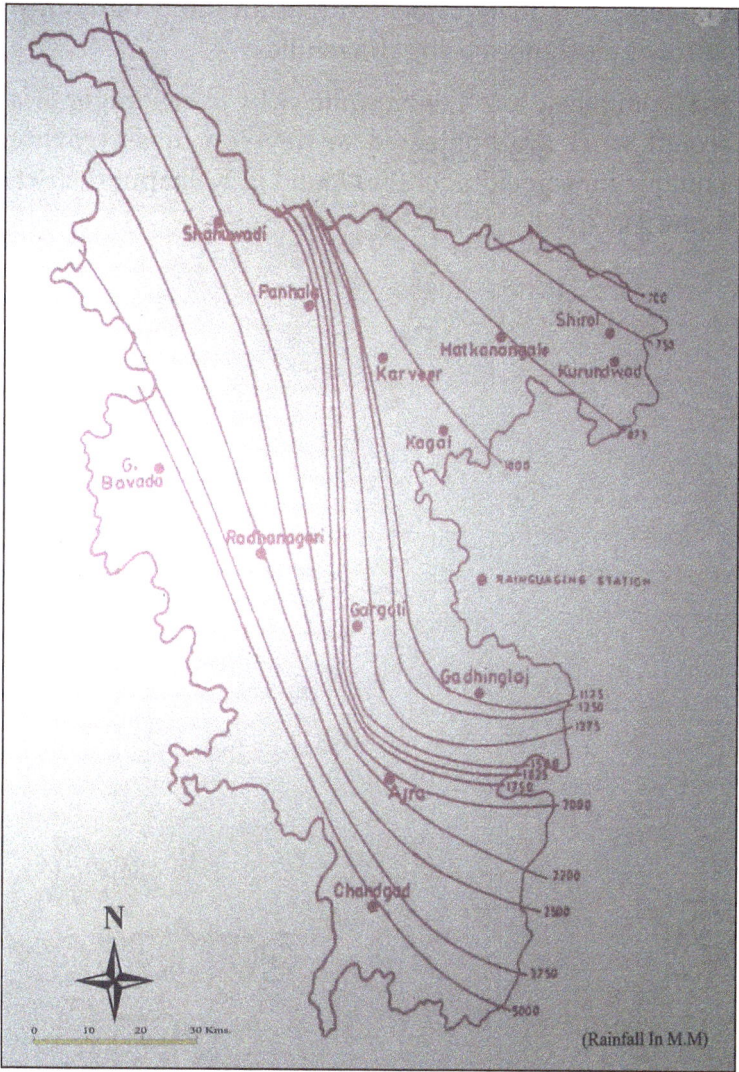

Figure 1: Map of Kolhapur district showing tahsils and rainfall.

1986; Sebastian *et al.,* 1990; Sathe and Girhe 2001, 2002; Sathe 2005; Sathe and Shinde 2008 etc.), attempted biodiversity of mosquitoes/dragonflies and biological control of mosqutoes using dragonflies.

Hoping the control of mosquitoes by using dragonflies present work was objected to develop mass rearing technique for some dragonflies found in Kolhapur district. (Figure 1).

Chapter 2
National and International Status

National Status

Although several species (550) of dragonflies are reported from India, (Fraser, 1933,1934,1936; Prasad 1995, 1996; Prasad and Varshney 1995; Prasad and Kulkarni; 2001,2002; Kulkarni and Prasad, 2002; Sathe and Shinde, 2008; Sathe and Bhusnar, 2011, etc.) no dragonflies are used for biological control of insect pests in India. The present work is first attempt from India. However, Sathe and Shinde (2008) attempted dragonflies with respect to biological control of paddy pests from Kolhapur region as primary trials.

International Status

Out side India, sufficient work has been reported on biological control of mosquitoes. In Cuba, Santhamarina and Mijares (1986) made field and laboratory observations to study efficacy of nymphs of Anisopterans *Pantala flaviscens* and *Tramea abdominalis* and found good predators of mosquito *Culex quinquefaciatus*. Thomson *et al.* (1988) studied predatory efficiency of nymphs of *Bradinopyga jaminata* and *Brachythemis contaminata* on mosquito larvae and found efficient biocontrol agents of mosquitoes at larval stage. A sizable work has been made from Japan in connection with the control of various types of mosquito larvae. Urabe *et al.* (1986) evaluated the predatory capacity and efficiency of *Sympetrum frequens* against the larvae of mosquito, *Anopheles sinensis* in the laboratory and reported that when the nymph size increased subsequently the number of mosquito larvae consumption also intensified. They also reported the predatory prey relation between *Sympetrum frequens* and the larvae of *Anopheles sinensis* in rice field near Omiya, Japan and revealed that the density of mosquito larvae increased when the nymphal density of the predator become low. The distribution patterns of the predators and prey in the smaller field had a non-overlapping tendency indicating effective predation.

A good example known about dragonflies have successfully through augmentative release (AR) was the introduction of half grown larvae of the libellylid, *Czocontenemis servelia* into domestic water storage containers in Yagong (Rangoon) in Myanmar (Burma). The water storage containers were used by the aquatic stage of the yellow fever mosquito *Aedes aegypti*, which was also responsible for the transmission of dengue fever in that locality. More than 92 per cent of the local population of *A. aegypti* was occupying the containers, which because of their function were easily accessible to householders and control operators. The systematic release of dragonfly larvae during the monsoon season (the time when dengue fever was being transmitted by the mosquito) rapidly depressed the mosquito population to a level lower than that could have been achived by any other known method, including treatment by chemical insecticide. Sebastian *et al.* (1990) demonstrated the trials of effectiveness of the approach. The Rangoon trial was very much successful, due to fact both the adult and larvae of dragonflies were effective for controlling the population of mosquitoes as like conventional pesticides. The present study would worth while utilizing dragonflies in biological control of mosquitoes.

Chapter 3
Project Profile

i. **File no.** : BT/PR9962/AGR/05/396/2007.

ii. **Title of the Project** : "Mass rearing of Dragonflies for Biological Control of Mosquitoes in Kolhapur district"

iii. **Date of Sanction** : June, 15th, 2009.

iv. **Duration of project** : 15 June, 2009 to 15 June, 2012.

v. **Name of principal investigator** : Prof. T. V. Sathe, Head, Department of Zoology, Shivaji University, Kolhapur

vi. **Objectives:**

Short term objectives:

a) To make survey of dragonflies in Kolhapur district.

b) To study life-cycle and fecundity of few species. (selected) of dragonflies.

Long term objectives

c) To develop mass rearing technique for some potential dragonflies as biocontrol agents of mosquitoes.

d) Objective recommended by task force committee: To study Ecology of mosquitoes.

Chapter 4
Methodology

DETAILS OF METHODOLOGY

i) Methodology for Survey and Surveillance of Dragonflies

Sampling insect population is one of the important tasks in population dynamics which estimate the number of the species present in the target area. For sampling the insects with specialized habits like predators require basic knowledge on biological peculiarities. In general, the predators are usually specific for one or closely related group of insects. Thus, they are restricted to, and dependent on an individual of prey species. Their reproductive rates are also related to the availability of their preys. Thus, predatory sampling is the process of sampling of the pest species. The predators frequently determine the population densities of the pest species. In general, the survey and surveillance of pest and their predators has tremendous economic importance in the pest management. Survey studies or estimation of number of predators per unit of land area therefore necessary to construct life table and for control release against pest species in the field conditions.

Survey of dragonflies has been made by visiting different study spots in Kolhapur district (Figure 1) namely, Kagal, Gadhinglaj, Panhala, Karveer, Shirol and Malkapur at 15 days interval and collecting adults (Figures 2-5)/ naiads (Figure 6) one man one hour search method. The naiads were collected from aquatic habitats with the help of aquatic net (Figure 7) and plastic containers (Figure 8) and adults with regular insect net (Figure 9) from terrestrial habitats associated with water bodies. Later, the collected species have been identified by consulting appropriate literature (Fraser, 1933, 36, 54; Sathe and Shinde,2008 etc.). Naiads were reared in glass aquarium size (3x2x1 ft.) (Figure 10) in the laboratory for adult emergence and for further experiments.

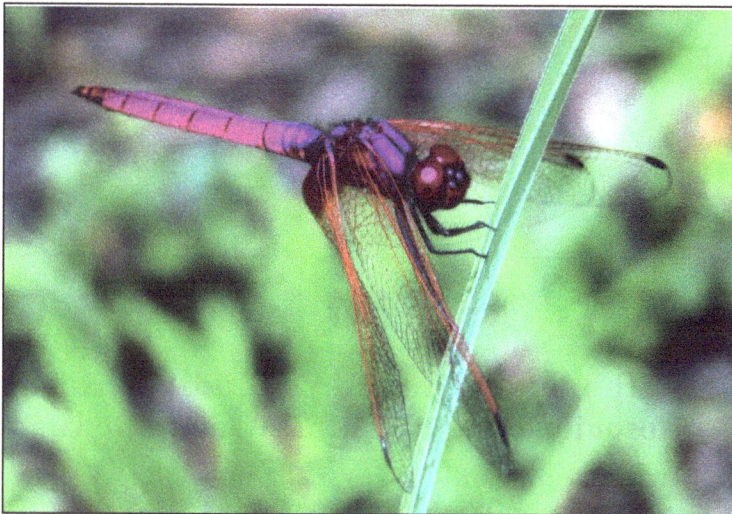

Figure 2: *C.S. servilia* (Adult).

Figure 3: *Diplacodes trivialis* (Adult).

Figure 4: *Pantala flaviscens* (Adult).

ii) Methodology to Study Biodiversity of Mosquitovorus Dragonflies (Order:Odonata)

Survey of dragonflies was made by visiting various Tahsils of Kolhapur district including Westerns Ghats at weekly interval from the years 2008 - 12. Mostly, spot

Figure 5: *Neurothemis tullia tullia* (Adult).

Figure 6: Naiads of Dragonfly *Pantala flaviscens*.

Figure 7: Aquatic insect net for collecting naiads.

Figure 8: Plastic container for collection of naiads.

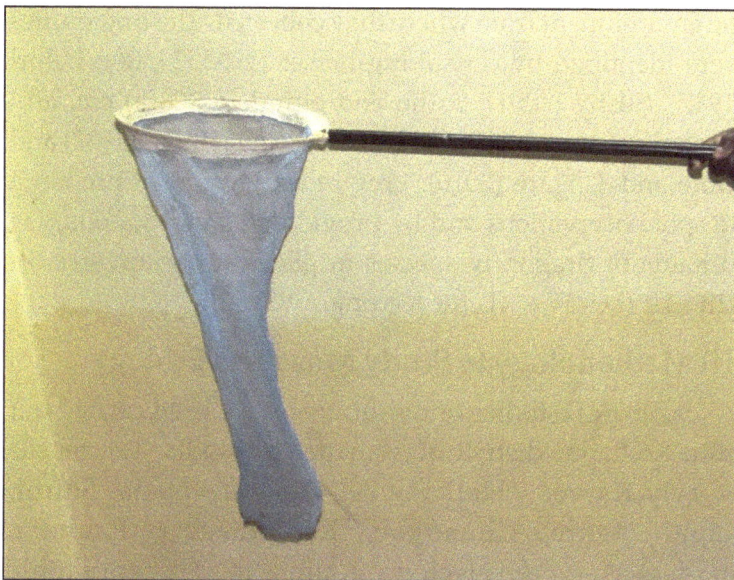

Figure 9: Insect net for collecting dragonfly adults.

Figure 10: Large glass aquarium (Size 3 × 2 × 1 feet).

observation was followed by photography and Odonates were collected from the Ghats for their identification. After noting characteristics the live dragonflies were released in the environment from where they collected. The dragonflies were identified by consulting Fraser (1933, 34, 36); Lahiri (1987), Silsby (2001), Sathe and Shinde (2008 a) etc. and mosquitoes were identified by consulting Baraud (1937) and Sathe and Tingare (2010). Prey index have been prepared by spot observations and by providing 50 mosquito larvae to naids of dragonfly species in glass Aquarium size 3ft x2ft x1ft (L × W × H) for feeding.

iii) Methodology to Study Mosquito Ecology

Ecological studies of mosquitoes have been carried out from Kolhapur district of Maharashtra, India. Ten tehsils namely Karveer, Hatkangale, Gaganbavawda, Shirol, Kagal, Azara, Chnadgad, Radhanagari, Panhala, Shahuwadi were selected on climatic and geographic features. Kolhapur district (Figure 1) is located between 15°C to 17°C, North latitude and 73°-74°, East latitude, containing 1,46,575 hectares land and uneven rain fall from 6000 mm at Gaganbawada to minimum 500 mm at Hathkanagle. Therefore, the district contains 2 major, 10 medium, and 31 minor water bodies, the sources of breeding mosquitoes.

The survey of mosquitoes was carried out during the years 2009- 2011 from Kolhapur above water bodies and tahasil by visiting 15 days interval by one man one hour search method specially, evening hours. The adult mosquitoes were collected with the help of suction tube larvae by ½ litre glass beaker. Random samples from breeding places both natural (small lakes, streams, ponds, mud, pools, river, free holes, etc.) and artificial (irrigations channels , wells, cement tanks, container type like plastic, metal cans, earthen ware pots, bamboo strips, etc.) were taken. Later, the larvae have been reared in the laboratory for adult emergence. The mosquitoes were identified consulting Barraud (1934), Christophers (1933), Rao (1984), Sathe and Shinde (2002) and Sathe and Tingare (2010). Environmental temperature, rainfall and relative humidity were monitored during the study.

iv) Methodology to Study Biology of a Dragonfly *Bradinopyga jaminata* (Rambur)

Last instar nymphs of *B. jaminata* have been collected from Shivaji University campus and reared in 1 Lit. plastic container by providing mosquito larvae/mud worms at laboratory conditions (27 ± 1°C, 80 per cent R.H and 14 hr photoperiod) for adult emergence. Newly emerged adults were isolated individually in 1 Lit plastic container and later, a pair (1 Male and 1 Female) was released into a large insect cage (size 25' × 15' × 10' in length, width, height) (Figure 11) fitted on a water tank (size 20 × 10 × 2.5, in length,

Figure 11: Large insect nylon net (Bamboo nylon net).

width, height) (Figure 12) for mating purpose for 24 hr.
After mating, female has withdrawn and allowed to lay
eggs in glass aquarium (size 3' × 2' × 1' in length, width,

Figure 12: Water tank of size 20 × 10 × 2.5 (length, width, height).

Figure 13: A: Small aquarium (45 × 22 × 28 cm);
B: Naiads of *C. servilia servilia*

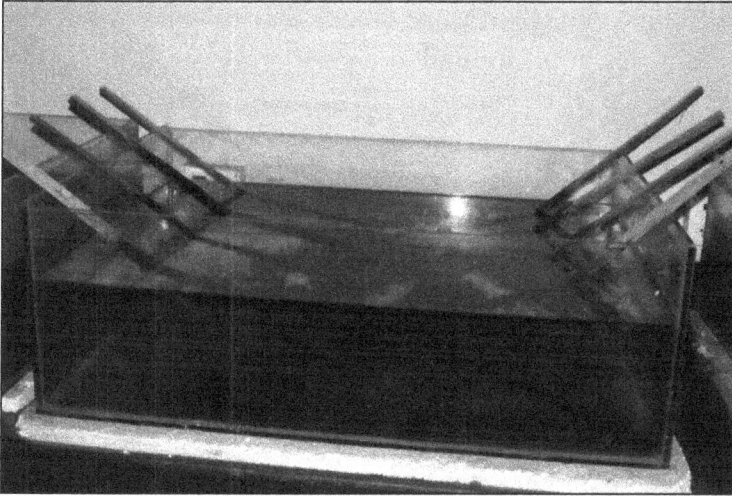

Figure 14: Aquarium with wooden naiads reeds.

height). After hatching the eggs, 1st instar naiads were fed with paramecium, 2nd with daphnia, 3-6 with *Aedes* mosquito larvae and 7 to 10 with mosquito larvae/mud worms/bee grubs and reared in 1 Lit. plastic container, small aquarium (45 × 22 × 28 cm in length, width height Figure 13) and large glass aquarium (3' × 2' × 1') respectively. For last instar wooden naiads reeds (Figure 14) have been provided to glass aquarium for climbing upon them and for attachment of exuviae for emergence of adult dragonfly. Thus, ten females have been studied for biological and fecundity studies. Offsprings emerged have been counted and sex ratio has been noted in all the cases.

v) Methodology to Study Biology of a Dragonfly *Crocothemis servilia servilia* Drury

Nymphs of *C. servilia servilia* have been collected from the water bodies of Shivaji University, campus and reard in laboratory (24 ± 1°C, 65-70 per cent RH and 2hr photo period) for adult formation on red worms and mosquito larvae. On emergence, adult males and females were separated and kept in rearing cage and identified consulting Fraser (1934 to 1936). Later, exposed for mating in nylon, bamboo net/cage (15 feet × 15 feet 10 feet in length, width and height respectively.) After mating females were allowed to lay their eggs in the water in plastic through of diameter 30cm. After hatching the eggs first instar nymphs and second instar nymphs where reared in the same plastic throughs by providing paramecium, daphnia etc. as food. Later, the nymphs (naiads) have been reared in aquarium 3 feet length × 2 feet width × 1 feet height by providing mosquito larvae and observations where taken on the nymphal (naiad) duration and later, adult longevity in the laboratory was studied. Sufficient number of individuals were reared for conformation of life cycle and fecundity studies. In control, no food was provided to adults. Fecundity was studied by exposing one mated female in nylon cage covered over water tank (Figure 11).

vi) Methodology to Study Adult Emergence Behaviour in *Crocothemis servilia servilia* Drury

C. s. servilia matured instars were collected from Shivaji University, Kolhapur water tank (Figure 15) and reared in large glass aquarium (90x 60 × 30cm) under laboratory conditions (27±1°C, 80 per cent R.H., 14 h photoperiod) by providing mud worms/mosquito larvae and honeybee grubs for adult emergence. In the aquarium, wooden reeds were placed for naiads for climbing and fixing body upon them for adult emergence. Observations on adult emergence were

Figure 15: Shivaji University, Kolhapur water tank.

noted every hour per day and once the emergence process initiated, entire thing was observed. The experiment was replicated ten times.

vii) Methodology to Study Mosquito Larvae Consumption Rate by a Dragonfly *Pantala flaviscens* (Rambur)

Dragonfly naiads of *P. flaviscens* have been collected and reared for adult emergence in the laboratory in a aquarium of size, 3' × 2' × 1', (length, width, height) containing a water tank of size, 15' × 10' × 2' (length, Width, depth) for mating purpose. The mated female has been allowed to lay eggs in the glass aquarium water. After hatching the eggs, the first and second instar nymphs have been reared on paramecium, Daphnia and red worms while, 3rd to 10th instars have been reared on second instar larvae of *Aedes* mosquito in above said aquarium. For each instar 100 larvae of mosquito were exposed for 24 hr and consumption rate of mosquito larvae by *P. flaviscens* (naiads) has been recorded.

viii) Methodology for Rearing of Paramecium

Initial culture of Paramecium was collected from Aquatic and muddy habitats from Shivaji University, Kolhapur water tanks and 20-25 paramecium were taken into 1 *l* plastic jar with the help of water dropper/ inoculation pipettes. The jar was closed with polythene gasket. Dried and decomposed folks and corn husk and a growth medium (containing NaCl–75mg, NaH CO_3–3mg, KCl–3mg, $CaCl_2$–3mg, $CaH_4(PO_4)$, H_2O–1.5mg, 10 drops of skimed milk in 1 lit. of distilled water) was used for culture of paramecium. In addition, a piece of hydrilla plant and 5-10 drops of skimmed milk was also added in the jar for providing required bacteria and small protozoans. The paramecium reproduced by simple division 2-3 times per day and can move 4 times body length per second. Thus, sufficient number of paramecia (Figure 16) are produced.

Three compartmental glass aquarium (Figure 17) of size 45 × 30 × 30 cm (each compartment 15 × 30 × 30 cm, in length, width and height respectively) was used for mass culture of *P. caudatum*. Three compartments were perforated by small quadrangular holes at a height of 7, 5 and 3 cm which facilitated the flow of paramecium from one compartment to another compartment periodically and constant removal of paramecia from the compartment. Half

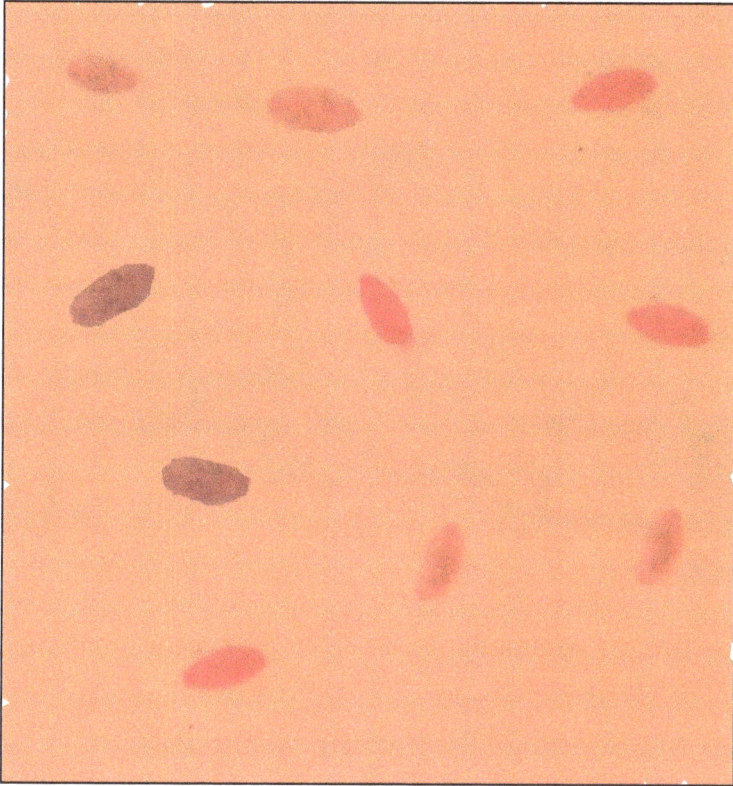

Figure 16: Paramacia produced under laboratory conditions.

lit of paramecium initial culture obtained from glass jar method was taken in 5 *l* of distilled water in glass aquarium equipped with sliding cap, an aerator, some hydrilla plants and paramecium food (500 g of folks and corn husk with equal proportion, 1 Parle-G Biscuit). The culture was allowed to reproduce for 1 week in the first compartment and then culture was allow to flow/release to second and third compartment or replaced and used for rearing of first instar naiads of dragonflies. After 1 week half litre of water was added daily with the appropriate food quantity for

Figure 17: Glass aquarium for mass culture of paramacia.

the flow of paramecia from one compartment to another compartment.

ix) Methodology for Rearing of Daphnia

Daphnia were collected from the natural habitats for starting initial culture with the help of aquatic net (Figure 18). In an aquarium (Figure 13) 45 × 22 × 28 cm (length, width, height) 5 lit of water was taken and 20-25 Daphnia were released with the help of water dropper. 2-3 snails for faecal matter, cabbage leaves as food for snail and some algae and rice bran were also added into the aquarium in addition to some small protozoans and rotifers. Daphnia (Figure 19) were allowed to multiply for about 6 weeks and huge number of Daphnia were collected and used for rearing second instar naiads of dragonflies. The experiment was conducted at laboratory conditions (27±1° C, 75±1 per cent R.H. and 12 hr photoperiod).

Figure 18: Aquatic net for collection of Daphnia.

x) Rearing Method of Mosquito Larvae

From aquatic habitats and drainage mosquito larvae have been collected and identified by consulting Smith (1993) Barraud (1934) and Rao (1981) and reared in 1 lit. plastic container for adult emergence. The adults were fed

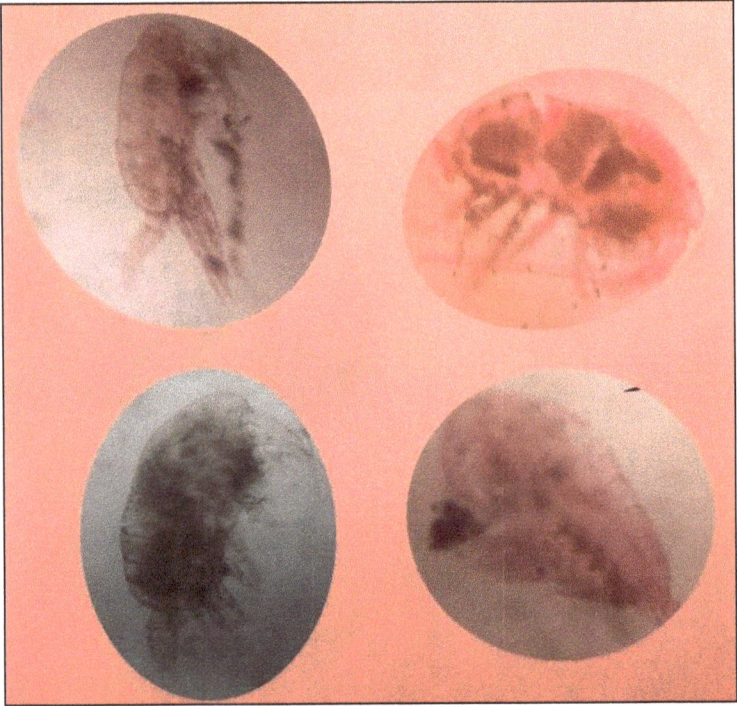

Figure 19: Daphnia for mass rearing.

with 20 per cent honey solution and/or allowed to bite rats/mouse in the insect cage (Figure 20) size 25 × 25 × 30 cm (in length, width and height) and then transferred to another similar insect rearing cage for mating. A petridish containing fresh water and moist filter paper was then kept in the insect rearing cage for 24 hr for oviposition of mosquitoes. The eggs were collected on moist paper and hatched 1st instar larvae were taken into 1 lit. plastic containers. 200-400 eggs are laid by mosquitoes. First to fourth instar larvae of mosquito were provided with rat faecal matter, mud, dried blood of goat and powdered liver 1 gm/lit. On an average 0.4 mg/larva food requirement is

Figure 20: Inesct rearing cage.

essential. Larval stage lasts for 4-9 days in *Aedes* mosquitoes, 1 week in *Anopheles* and 7-10 days in *Culex*. Second to fourth instar larvae of mosquito were reared in the glass aquarium size, 3 ft × 2 ft × 1 ft (in length, width and height) and glass aquarium with tin cap (Figure 21). The pupae were reared in the same glass aquarium for 1 day. Later, transferred to plastic containers (5 × 7 × 2.5 cm) for adult emergence or 2000 pupae per insect cage (25 × 25 × 30 cm). Thus, sufficient number of adults and larvae (Figure 22) have been collected.

Second Method of Mosquito Rearing

Troughs and mud pots of size 3 litre capacity have been placed with water outside the animal house of rats and

Figure 21: Mosquito rearing cage.

Figure 22: Mosquito larvae.

allowed mosquito species to lay eggs in the pots. After egg laying of mosquitoes in the above pots sufficient number of mosquito larvae have been collected and were identified, mostly as *Aedes* sp., have been used as a food for naiads in mass rearing technique at scarcity of food for dragonflies.

xi) Methodology for Rearing of Dragonflies in Large Glass Aquarium

Matured naiads of *Crocothemis servilia servilia* Drury have been collected from Shivaji University campus (tanks) and reared in glass aquariums (size 3 ft × 2 ft × 1 ft) (Figure 23) at laboratory conditions (27±1°C, 80 per cent R.H. and 14 hr photoperiod) individually. Newly emerged adults were kept individually in 1 lit plastic container for some time and then one male and one female (Figure 24) released in the large insect net 25' × 15 ' × 10' (in Length, width and height) (Figure 11) fitted in the rearing water tank of size 20 × 10 × 2.5 ft (in L, W, H) (Figure 11) for mating purpose for 24 hr. Then female has been isolated and allowed to lay eggs in water in petridish (Figure 25) by deeping abdomen. Sufficient number of eggs have been collected on wet filter

Figure 23: Large glass aquarium for dragonfly rearing.

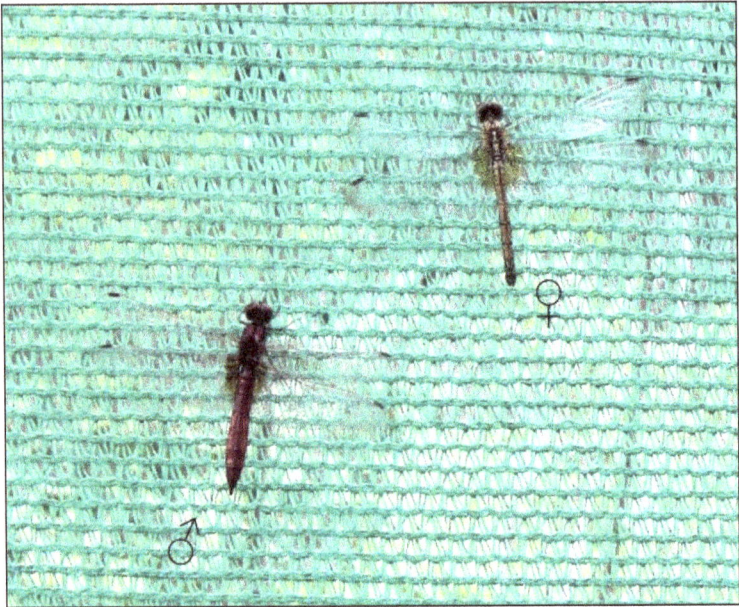

Figure 24: *Crocothemis servilia servilia* Male and Female.

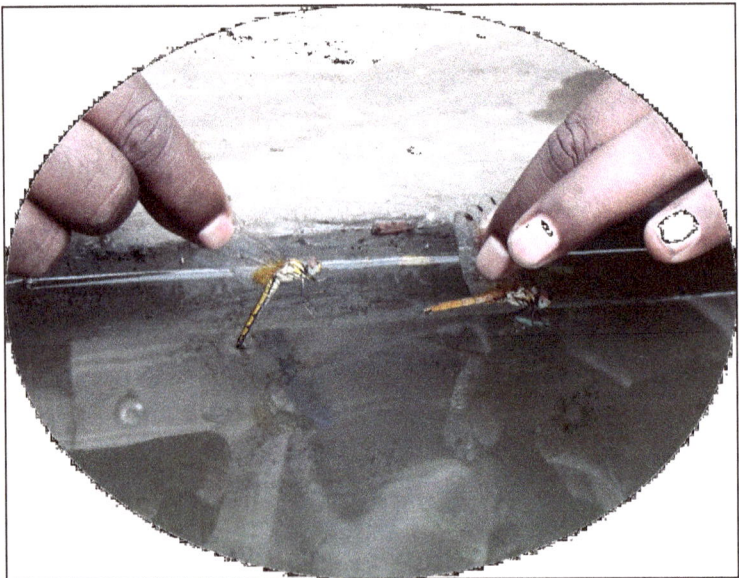

Figure 25: Oviposition by dragonfly.

paper. Eggs hatched in about 5-8 days. The first instar naiads are very small sized and light coloured. They are taken in 1 lit plastic container for rearing purpose. For first instars paramecium is best food. Paramecium can be cultured as per the method given earlier. The plastic container (Figure 26) should contain little duck weed/ Hydrilla for aeration. Within 2-3 days first instars moult into second stage (instars). The second instars are now released into a glass aquarium of size $45 \times 22 \times 48$ cm (L,W,H) and provided Daphnia (Figure 19) as food in addition to some Hydrilla plants for aeration. This stage lasts for 3-5 days. Then third instar onward naiads are transformed to large glass aquarium (size $3 \times 2 \times 1$ ft (in L, W, H). 3rd to 6th instars are given mosquito larvae (Figure 22) for feeding purpose, provided either mosquito larvae/Mud worms/ bee grubs as food. There are 12 instars in *C. serillia servilia.*

Figure 26: Plastic container.

For last instar, wooden reeds (Figure 14) are fixed in the glass aquarium so that the naiads can climb upon them for fixing body and casting exuviae and for adult emergence. Adult emergence process was completed within 2 to 3 hours mostly from, 3.00 a.m. to 6.00 a.m. or sometimes at morning hr. Sufficient number of naiads and adult dragonflies can be had by this method.

xii) Methodology for Rearing of Dragonflies in Water Tank Covered Net

Newly emerged pair (male and female) of *Crocothemis* are released in the water tank covered with net size, (Figure 27) (25 × 20 × 10 ft) for mating for 24 hr. Later 5 to 10 mated females be released in a cage containing water tank (size 20 × 15 × 2.5 ft. (in L,W,H) for 24 hr for egg laying. The females will start egg laying in the tank water. Thus, sufficient number of eggs will be received for further rearing of naiads. The tank should contain some plants of duck weed/Hydrilla (Figure 28) for aeration for naiads. Paramecium, Daphnia culture be made available at the bottom of tank, for first two instars and *Aedes* mosquito larvae culture for 3-6 instars. Then after mosquito larvae mud worms/bee grubs/red worms, be given to instars 7 to 12 as per the need. Final instar naiads will climb on the walls of tank for attachment of their exuviae to wall for facilitating adult emergence. Within 2-3 hr emergence process is completed and dragonfly will ready to fly after spreading the wings and stretching legs and body at morning. Thus, we can collect large number of naiads in the tank water and adults in the large insect cage.

Figure 27: Rearing nylon cages of size 25 × 20 × 10 feet on water tanks.

Figure 28: Water tank containing Hydrilla.

Rearing of *Bradinopyga jaminata* and *C. s. servilia*

The basic requirement was same except the duration of life cycle of the species.

		B. jaminata	C. s. servilia
Eggs laying	-	200-250	180–200
Incubation period	-	5-8 days	10–12 days
Larval period	-	45-54 days	47-56 days
Total instars	-	10	12
1st instar	-	2–3	2–3
2nd instar	-	3–5	3–4
3rd instar	-	5–6	5–6
4th instar	-	5–7	5–7
5th instar	-	5–7	5–7
6th instar	-	5–7	5–7
7th instar	-	5–7	5–6
8th instar	-	4–6	5–6
9th instar	-	4–5	3–5
10th instar	-	2–3	3–5
11th instar	-	0–0	3–5
12th instar	-	0–0	2–3
Progeny production/female (Fecundity)		188	176

xiii) Methodology for Mass Rearing Technique for Dragonflies

Methodology is given under the section experimental results.

Chapter 5
Experimental Results

i)   Survey and Surveillance of Dragonflies

Kolhapur district of Maharashtra, India contain several small and large water bodies as natural and artificial breeding places for mosquitoes. The distribution of various mosquito genera in natural and artificial habitats and their relative species abundance was studied during the years 2009-2012. In all, 22 mosquito species belonging to three genera *viz. Anopheles, Culex* and *Aedes*, which were potential vectors of malaria, dengue, chikungunia, filaria and Japanese encephalitis of human in the district surveyed. A total of 2113 mosquito larva were collected from artificial sources and 1993 larvae were harvested from natural sources. *Anopheles (Cellia) culicifacies* Giles, *Culex (Culex) fatigans* Widemam, *Aedes (Stegomyia) aegypti* Linnaeus and *Ae (S.) albipictus* Skuso were dominant over other species. Favorable environmental conditions for breeding mosquitoes and impact of human activities on population dynamics of mosquitoes have also been studied.

ii) Diversity of Mosquitovorus Dragonflies

The results recorded in Table 1 and Figure 29 shows that 43 species of mosquitovorus dragonflies were common in Kolhapur district including Western Ghats, Maharashtra.

Augmentative release of biocontrol agents is practiced routinely in several countries for the suppression of mosquitoes (Corbet, 1999). Thomas *et al.* (1988) studied efficacy of *Bradinopyga jaminata* and *Brachythemis contaminata* against mosquito larvae while, Santamarin and Mijares (1986) studied the mosquito larvae feeding potential of *Pantala flaviscens* and *Tramea abdominalis*. They tested *Culex quinquefaciatus* larvae against above two dragonflies and found that above two dragonflies are good biocontrol agents of mosquito *C. quinquefaciatus*. The larvae of *Anopheles sinesis* have also been tested against *Sympetrum frequens* (Urabe *et al.*, 1986). Very interestingly *Aedes aegypti* mosquito population was suppressed upto 92 per cent in Myanmar by using a libellulid dragonfly *Czocontenemis servilia* (Sebustian *et al.*, 1990).

Recently, Sathe and Shinde (2008 b) prepared a index of dragonflies and their prey mosquitoes from Orus region of Sindhudurg district of Maharashtra (Western Ghats). Sathe and Shinde (2008 b) also demonstrated the use of dragonflies in insect pest management specially mosquitoes. The present work will add great relevance in designing mosquito biological control programme by using dragonflies.

Table 1: Biodiversity occurrence and mosquito preys of dragonflies.

Sl.No.	Dragonfly Species	Sub-family	Prey species	Occurrence
FAMILY – GOMPHIDAE				
1.	*Gomphus* sp.	Gomphinae	*Mosquitoes*	Common
2.	*Gomphus nilgiricus* (Laidlaw)	Gomphinae	*Anopheles* sp.	Common
3.	*Burmagomphus laidlaw* (Fraser)	Gomphinae	*Culex* sp.	Common
4.	*Burmagomphus pyramidalis* (Fraser)	Gomphinae	*Culex* sp.	Common
5.	*Davidioides martini* (Fraser)	Gomphinae	*Anopheles* sp.	Common
6.	*Mesogomphus lineatus* (Selys)	Gomphinae	*Aedes* sp. *Anopheles* sp.	Rare
7.	*Lamelligomphus malbarensis* (Fraser)	Gomphinae	*Culex* sp.	Common
8.	*Lamelligomphus nilgiricus* (Fraser)	Gomphinae	*Aedes* sp.	Rare
9.	*Megalogomphus superbus* (Fraser)	Gomphinae	*Culex* sp.	Rare
10.	*Cyclogomphus ypsilon* (Selys)	Gomphinae	*Culex* sp.	Common
11.	*Cyclogomphus heterostylus* (Selys)	Gomphinae	*Culex* sp. (Sept. oct)	Common
12.	*Merogomphus longistigma tamaracherriensis* (Fraser)	Gomphinae	*Culex* sp.	Rare
13.	*Onychogompus striatus* (Fraser)	Gomphinae	*Culex* sp.	Rare

Contd...

Table 1–*Contd...*

Sl.No.	Dragonfly Species	Sub-family	Prey species	Occurrence
14.	*Acrogomphus* sp.	Epigomphinae	*Culex* sp.	Common
15.	*Microgomphus* sp.	Epigomphinae	*Anopheles* sp.	Common
16.	*Microgomphus longistigma* (Fraser)	Epigomphinae	*Anopheles* sp.	Rare
17.	*Heliogomphus* sp.	Epigomphinae	*Aedes* sp.	Common
18.	*Macrogomphus annulatus* (Selys)	Epigomphinae	*Culex* sp.	Common
19.	*Macrogomphus wynaadicus* (Fraser)	Epigomphinae	*Culex* sp. *Aaedes* sp.	Common
20.	*Ictinus* sp.	Inctinae	*Aedes* sp.	Rare
FAMILY – CORDULEGASTERIDAE				
21.	*Chlorogomphus xanthoptera* (Fraser)	Chlorogomphinae	*Culex* sp.	Common
22.	*Chlorogomphus campioni* (Fraser)	Chlorogomphinae	*Culex* sp.	Common
FAMILY – AESHNIDAE				
23.	*Anax parthenope* (Selys)	Anaxinae	*Aedes* sp. *Culex sp.*	Rare
24.	*Anax immaculifrons* (Rambur)	Anaxinae	*Aedes* sp.	Common
25.	*Gynacantha basiguttata* (Selys)	Aeshinae	*Aedes* sp.	Common

Contd...

Table 1–*Contd...*

Sl.No.	Dragonfly Species	Sub-family	Prey species	Occurrence
26.	*Gynacantha millardi* (Fraser)	Aeshinae	*Anopheles* sp.	Common
FAMILY – LIBELLULIDAE				
27.	*Macromia indica* (Fraser)	Cordullinae	*Anopheles* sp. *Culex* sp.	Common
28.	*Macromia flavincia* (Selys)	Cordullinae	*Anopheles* sp. *Culex* sp. *Aedes* sp.	Rare
29.	*Macromia flavovittata* (Fraser)	Cordullinae	*Anopheles* sp. *Culex* sp. *Aedes* sp.	Rare
30.	*M. cingulata* (Rambur)	Cordullinae	*Anopheles* sp. *Aedes* sp. *Culex* sp.	Common
31.	*Epophthalmia vittata vittata* (Burmeister)	Cordullinae	*Anopheles* sp.	Common
32.	*Epophthalmia frontalis binocellata* (Fraser)	Cordullinae	*Anopheles* sp. *Aedes* sp.	Common
33.	*Macromidia* sp.	Cordullinae	*Aedes* sp.	Common
34.	*Idionyx optata* (Selys)	Cordullinae	*Culex* sp.	Common

Contd...

Table 1–*Contd...*

Sl.No.	Dragonfly Species	Sub-family	Prey species	Occurrence
35.	*Hemicordulia asiatica* (Selys)	Corduliinae	*Anopheles* sp. *Culex* sp.	Common
36.	*Amphithemis mariae* (Laidlaw)	Libellulinae	*Culex* sp.	Rare
37.	*Hylaeothemis fruhstorferi* (Kirsch)	Libellulinae	*Aedes* sp.	Common
38.	*Potamarcha obscura* (Rambar)	Libellulinae	*Culex* sp. *Aedes* sp.	Common
39.	*Orthetrum sabina* (Drury)	Libellulinae	*Culex* sp.	Common
40.	*Pantala flaviscens* (Rambur)	Libellulinae	*Anopheles* sp. *Culex* sp.	Common
41.	*Sympetrum* sp.	Libellulinae	*Anopheles* sp.	Rare
42.	*Crocothemis servilia servilia* (Drury)	Libellulinae	*Anopheles* sp. *Aedes* sp. *Culex* sp.	Common
43.	*Bradinopyga jaminata* (Rambur)	Libellulinae	*Anopheles* sp. *Culex* sp. *Aedes* sp.	Common

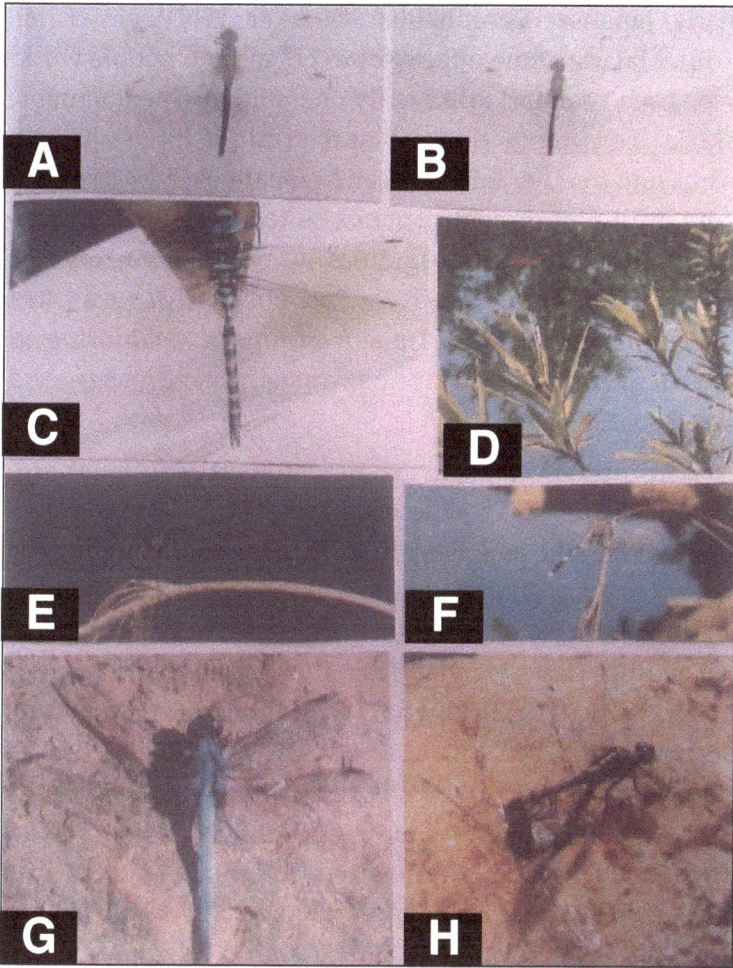

Figure 29: Biodiversity of dragonflies from Kolhapur district.
A: *Trithemis festiva*; B: *Trithemis arora*: C: *Anax immaculifomns*:
D: *Crocothemis servilia servilia*; E: *Bradinipyga* sp.; F: Unidentified;
G: Unidentified; H: Dragonflies-mating.

iii) Mosquito Ecology

Mosquitoes cause a huge medical and financial burden by spreading malaria, yellow fever, dengue, chikungunya, filaria, Japanese Encephalitis, RV fever, WNV, etc. (Fang, 2010). Malaria alone infects some 247 million people world wide each year and kills nearly one million. The uncurable filaria, confusing dengue and fatal JE have showed increasing trend of infections and casualities in India (Sathe *et al.,* 2010). Hence, control of mosquitoes is control of above diseases. Control of mosquitoes by pesticides is chronic problem since mosquitoes have developed resistance against most of the pesticides of major groups *viz.,* Chlorinated hydrocarbons, organophosphorus, carbamates and pyrethroids.

Mosquitoes can live on almost every continent and habitat. There are 3,500 described species of mosquitoes in the world (Sathe and Tingare, 2010). Therefore, hoping the biological control of mosquitoes through natural enemies, ecological studies of mosquitoes have been objected. Review of literature indicates that Bates (1949), Causey and Santos (1949), Service (1963), Okorie (1978), Guimaraes *et al.* (2000), Godwin *et al.* (2005), Silver (2008), Fang (2010) etc. worked on ecology of mosquitoes from abroad while, from India Christophers (1933), Barrauds (1934), Rao (1984), Tewari *et al.* (1987), Nagpal and Sharma (1995), Sathe and Girhe (2001, 2002), Sathe and Jagtap (2008, 2009, 2010), Jagtap and Sathe (2008), Jagtap *et al.* (2008, 2010), Sathe and Bhusnar (2010) and Sathe *et al.* (2010) etc. contributed on biodiversity and ecology of mosquitoes.

The results recorded in Tables 2 and 3 indicates that 22 species of mosquitoes belonging to the genera *Anopheles, Culex* and *Aedes* which were potential vectors of Malaria, Dengue, Chikungunya, Filaria and JE of human in the district surveyed. The most prevalent species in the district were *Anopheles (Cellia) culicifacies, An. (C.) stephensi, An. (C.) subpictus, Culex (Culex) fatigans, Aedes (Stegomyia) albopictus* and *A. (S.) aegypti.* Other prominent mosquito species with adult and larval catches are shown in Table 2. Population of *Anopheles* mosquitoes was abundant in wet months leading to malaria cases. From November to March, dry months, *Aedes* and *Culex* mosquitoes were more prevalent leading to Dengue, Chikungunya, JE and Filaria in the region. The combination of favorable environmental temperatures, rainfall and high relative humidity were responsible for higher number of mosquito larvae harvestation at onset of rainfall and biggest in July-August and gradual fall in October. *C. (C.) fatigans,* which is vector of bancroftian filariasis is quite abundant in most of the tehsils of Kolhapur. Similarly, vector of JE *C. (c.) albopictus* and *Ae. (S.) vittatus,* vectors of alboviruses in general were also dominant in Kolhapur district.

Mosquitoes related to natural breeding sites in the study area specially tree and rock holes were mainly *Ae. (S.) vittatus* and *Ae. (S). aegypti.* In the breeding places such as rock holes, pools, leaves talks, plants and animal shells, tyres, plastic containers and discarded house materials were mostly *Anopheles* species. Terraces, drinking water containers, tyres, unused plastic container or broken containers, earthen pots, etc. harvested mainly *Aedes* species. Plastic wares had a significant high breeding of *Aedes* and *Anopheles* than of *Culex. Culex* can breed in dirty waters through out the year but slowed down in hot months. The seasonal abundance of mosquitoes is shown in Table 2. Impact of human activities such as carelessness of disposal of household plastic or earthen pot wastes, open containers, keeping tyres on terraces or near human habitation, leakage in drinking water system, leakages/ cratis to latrine, open latrine pipes, drainagaes, greatly facilitiated mosquitoes in the district. Creations such as dams, ponds, channels, wells, cement water tanks etc. also widely stimulated mosquito populations and provided endless scope to mosquitoes to breed in the open environment.

The mean monthly temperatures ranged between 17°C to 40°C in Kolhapur district. The lowest temperature were observed between the months of December to January (7°C to 15°C). The lowest rainfall occurred between the months January to March with a peak in July-August. *Ae.indica* the largest *Aedes* mosquito in the world (Sathe and Girhe, 2002) has been reported in a very heavy and continuous rains in Kolhapur. Very high rainfall (5000 mm) in Chandgad and Gaganbawada yielded *An. (An.) barbirostris* in higher number than other tahasils. However, *An. (An.) stephensi* yielded higher number of mosquitoes in Radhanagari and Panhala, having moderate rain suggesting plenty artificial and natural sources for mosquito breeding in the tahasils. Some species were mostly found in rainy season which includes, *C. (M), malayi, C. (B.) modestus, Ae. (S.) uniliniatus* and *An.(C.) theobaldi*. Therefore, it is advised that the contribution of human activities and increasing environmental modification to the breeding of mosquitoes should be minimized and selective vector (mosquito) control measures including mosquito larvae be initiated before onset of rainy reason. The control measures may be biological, use of Guppy and naiads of dragonflies or pesticides.

Guimaraes *et al.* (2000) studied the ecology of mosquitoes in areas of Serra do Mar State Park, State of Sao Paulo, Brazil. They systematized monthly human bait collections for three times a day, for periods of 2 or 3 h each, in sylvatic and rural areas for 24 consecutive months (1991-92). A total of 24,943 specimens of adult mosquitoes belonging to 57 species were collected during 622 collective periods. *Coquillettidia chrysonotum* was the most frequent collected mosquito (45.8 per cent) followed by *Aedes serratus* (6.8 per cent), *Cq. venezuelensis* (6.5 per cent), *Psorophora ferox* (5.2 per cent) and *Ps. albipes* (3.1 per cent). They further reported that temperature and rainfall influenced the inicidence of the mosquito species namely, *An. cruzii, An. mediopunctatus, Ae. scapularis, Ae. fulvus, Cq. chrysonotum, Cq. venezuelensis, Runchomyia reversa, Wyeomyia dyari, Wy. confusa, Wy. shanoni, Wy. theobaldi* and *Limatus flavisetosus* influenced by temperature change and relative humidity have effect on incidence of *Ae. serratus, Ae.scapularis, Cq.venezuelens* and *Ru. reversa* while, rainfall greatly influenced the population dynamics of the species, *An.cruzii, Ae.scapularis, Ae.fulvus, Cq. venezuelensis, Ru. reversa, Wy. theobaldi* and *Li. flavisetosus.*

Godwin *et al.* (2005) studied the ecology of mosquitoes of Midwestern Nigera by sampling mosquitoes by the method of Hopkins (1952), by dipping pipette or ladle. Their sampling sites were both natural and artificial. They also accounted climatic parameters like temperature, relative humidity and rainfall for studying incidence of mosquitoes. Their results revealed that 17 mosquito species belonging to genera *Anopheles, Culex* and *Aedes* were potential vectors of four human diseases. They encounted a total 736 mosquito larvae from artificial sources and 568 larvae from natural sources by pipette dipping collection method. In the present study comparatively a quite large number of mosquitoes have been collected by ½ litre beaker random sampling method. Godwin *et al.* (2005) further stated that pools, plastics and metal cans were the predominant artificial sources of mosquito larvae. More or less same situation is observed in the Kolhapur district. The probable reason for this might be the modified style of life of Indians.

Mosquito control by pesticides is chronic problem since mosquitoes have developed resistance to most of the compounds belonging to chlorinated hydrocarbons, organophosphorus, carbamates and pyrethroids. Therefore, biological control as ecofriendly is good option for mosquito control (Sathe, 2006). Mosquitoes are delectable things to eat and they are easy to catch. In the absence of mosquito larvae hundreds of species of fish would have to change their diet to survive. Many species of insect, spider, salamander, lizard, frog have primary food as mosquitoes. Bat feed on mosquitoes containing its gut about 2 per cent diet of mosquitoes.

Mosquito larvae feed on decaying leaves, organic detritus and micro-organisms. Therefore, they have associations with popular flora and fauna. Mosquito larvae of *Wyeomyia smthii* are important members of tight-knit communities in the 25-100 millilitre pools inside pitcher plant *Sarracenia purpurea* in North America. They live with a midge *Metriocnemus knabi* rotifers, bacteria and protozoa. According to Fang (2010) eliminating mosquitoes might affect plant growth. When other insects drown in the water, the midges chew up their carcasse and the mosquito larvae feed on the waste products, making nutrients such as nitrogen available for the plant. Within pitcher plants, protozoan diversity was more in the presence of mosquito larvae (Addicott, 1974). Without mosquitoes, thousands of plant species would lose a group of pollinators.

Eliminating mosquitoes would temporarily relieve human suffering. The efforts to eradicate one vector species would be futile, as its niche would quickly be filled by another. *Aedes aegypti* from scarp yards in Florida replaced by Asian tiger mosquito *Aedes albopictus* (Fang, 2010).

A survey on incidence of mosquitoes from Kolhapur region was carried out by Sathe and Girhe (2002). They noted *Culex pipiens* as dominant species in the Kolhapur region probably it might now be replaced by *Culex fatigans*. Since it was more abundant than other *Culex* species. Sathe and Girhe (2001) reported *Ae. dorsalis* during the years 2000-2001. However, during the current survey it was not found in the region. That might be replaced by some other species.

iv) Life Cycle, Fecundity and Biology of Dragonfly *Bradinopyga jaminata* (Rambur)

B.jaminata showed 3 stage of its life *i.e.* Egg (Figure 30) naiads (Figures 31-32) and adult (Figure 33). Eggs are rounded with bluish tinge. They measures 1.8mm in length and 1.6mm in width. Incubation period ranged from 5-8 days (average 6.5 days): Nymphal period ranged from 40 to 54 days. Ist, second, third, fourth, fifth, sixth, eighth, ninth and tenth instar ranged 2-3, 3-5, 5-6, 5-7, 5-7, 5-7, 4-6, 4-5 and 2-3 days respectively and progeny production averaged 182.4 individuals per female per generation (range - 170 - 192) with sex ratio (male: female) 1:0.83. Adult longevity averaged 7.2 days (range 6-8 days) and oviposition days 3.5 (range 3-5 days) (Table 2).

The dragonflies belong to the genus *Lestes* have shortest duration of their life cycle (about 3 months) and larger dragonflies have longest duration (1 year to 3 years) (Kumar, 1972c). According to Corbet (1980) the life cycle of

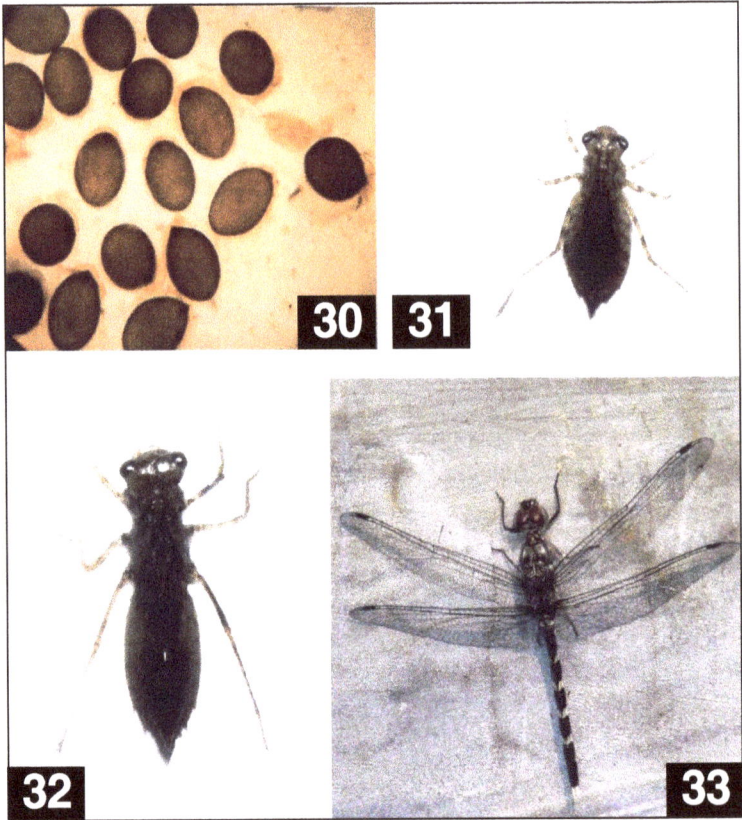

Figures 30–33: *B. jaminata.*
Figure 30: Eggs; **Figure 31**: 3rd instar naiad; **Figure 32**: 10th instar naiad; **Figure 33**: Adult.

dragonflies is amphibiotic and have 3 stages of life *viz.,* egg, numph (naiad) and adult. The adults are terrestrial but naiads are aquatic. According to Sathe and Shinde (2008) the nymphal duration of dragonflies vary greatly. Kakkassery (2004) says that robust naiads show labial mask for catching preys and rectal gills for respiration. According to Kumar and Prasad (1978) incubation period of egg varies from 5-40 to 80-230 days in tropical and temperate region

Table 2: Longevity, fecundity and sex ratio of *B. jaminata*.

Sl.No.	Replicates	Adult Longevity of Females	Oviposition Days	Progeny Production			Sex Ratio Male : Female
				Male	Female	Total	
1.	A	8	4	100	88	188	1 : 0.88
2.	B	7	3	98	80	178	1 : 0.81
3.	C	7	3	90	80	170	1 : 0.88
4.	D	6	3	92	81	173	1 : 0.88
5.	E	8	4	99	91	190	1 : 0.91
6.	F	8	4	99	89	188	1 : 0.89
7.	G	7	3	110	70	180	1 : 0.63
8.	H	8	4	112	80	192	1 : 0.71
9.	I	7	4	100	91	190	1 : 0.91
10.	J	6	3	93	82	175	1 : 0.88
Average		7.2	3.5	99.3	83.2	182.4	1 : 0.838

respectively. In *N. chinensis* Kumar (1973) counted 830 eggs. While in the present form maximum 250 eggs were laid. *Anax immaculifrons* lay eggs endophytically on submerged leaves while *Ictinogomphus rapax* lay eggs on tightly coiled filament for avoiding swepting down of eggs into stream (Kulkarni *et al.*, 1999). Likely, *Burmagomphus sivalinkensis* deposit eggs on the gelatinous material (Kumar, 1978). The present form lay eggs on water surface by dipping abdomen into water.

Kakkassery (2004) studied the food requirement for naiads of dragonflies and reported that small insects, fishes, tadpoles and redworms are good food material for naiads. He also reported pre reproductive period as few days to two to three months. In some dragonfly species colour change to body has taken places within the same sex or between sex however, males are more colourful than females (Corbet, 1980). In *B. jaminata* colour change found within the unmatured and matured individuals and the naiads can found feeding upon paramecium, daphania, *Aedes* mosquito larvae, mudworms and honey bee grubs.

Conclusion

1. *B. jaminata* is rearable in freshwater.
2. It has good potential to control *Aedes aegypti* masquito larvae.
3. Need mass rearing and utilization of *B. jaminata* for control of *A. aegypti* mosquitoes.

v) Life Cycle Fecundity and Biology of a Dragaonfly *Crocothemis servilia servilia*

The life history of this species is amphibiotic and has 3 stages *viz.* egg (Figure 36), naiad (nymph) (Figures 37-38)

Figure 34: *Crocothemis servilia servilia* (Male).

Figure 35: *Crocothemis servilia servilia* (Female).

and adult (Figure 34-35). Egg is rounded and bluish in colour, about 1.5 mm long and hatched within 15 to 18

Figure 36: Eggs – *C.s. servilia* .

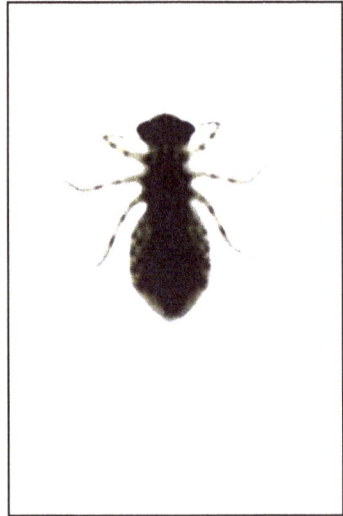

Figure 37: *C.s. servilia* First instar Naiad.

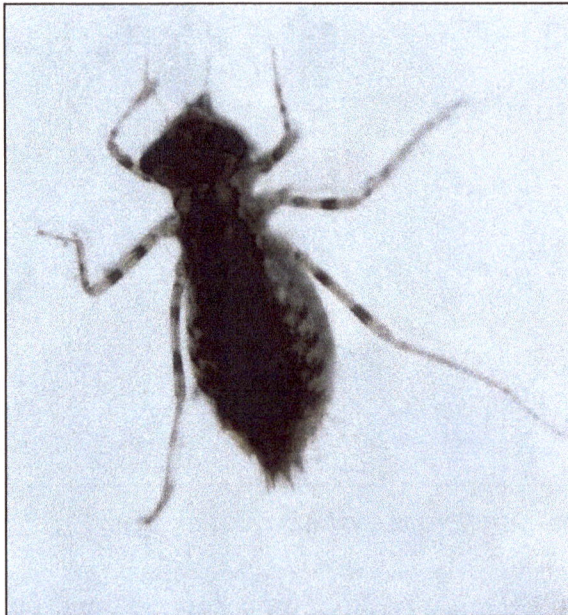

Figure 38: *C.s. servilia* Last instar Naiad.

Table 3: Longevity, Fecundity and Sex ratio of *C. servilia servilia*.

Sl.No.	Replicates	Adult Longevity of Females	Oviposition Days	Progeny Production			Sex Ratio Male : Female
				Male	Female	Total	
1.	A	7	3	100	76	176	1 : 0.76
2.	B	7	3	98	80	178	1 : 0.81
3.	C	7	3	109	82	191	1 : 0.75
4.	D	6	3	90	80	170	1 : 0.88
5.	E	7	3	91	82	173	1 : 0.90
6.	F	6	3	99	89	188	1 : 0.89
7.	G	6	2	93	80	173	1 : 0.86
8.	H	6	2	90	80	170	1 : 0.88
9.	I	6	3	94	82	175	1 : 0.86
10.	J	7	3	111	78	189	1 : 0.70
Average		6.5	2.8	97.5	80.8	178.3	1 : 0.829

days. Naiad is flat bodied, measuring about 2 cm in body length. Last instar naiad shows labial mask for catching preys. It also shows rectal gills in abdomen. There are 12 instars, each has about 7-10 days duration. During nymphal period they feed on paramecium, red-worm, daphnia and mosquito larvae. Adults survived for 4 days without food. Single mated female laid 140-150 eggs in water body/water trough and produced on an average 142 adults under laboratory conditions. Fecundity was studied by exposing a pair (one male, one female) in nylon cage (Figure 27). The adult longevity averaged 6.5 days (range 6 -7 days) and oviposition days 2.8 (2-3 days), the progeny production averaged 178.3 (range 170-191) with sex ratio (m:f) 1:0.82 (Table 3).

According to Sathe and Shinde (2008) the aquatic life of larvae varies greatly in length. The dragonflies belonging to the genus *Lestes* have shortest duration of about three months while, the larger dragonflies showed longer period, about one year to 3 years. According to them, "The life history of dragonfly is amphibiotic and has 3 stages, *viz.* egg, nymph (naiad) and adult." The larval stage is passed in watery environment and adults are terestrial. In the present form *C.s. servilia* same situation is recorded regarding life stages and habitat *i.e.* life cycle contains egg, nymph (naiad) and adult and the individual is amphibiotic. According to Kakkassery (2004) robust naiads shows labial mask for catching prey, rectal gills are within abdomen. Similar modifications are noticed in *C.s. servilia*.

Kumar and Prasad (1977) counted 830 eggs in *N. chinensis* in 55cm² leaf area of fern plant. According to Kumar and Prasad (1977) incubation period of egg varies from 5-40 to 80-230 days in tropical and temperature regions. *Anax immaculifrons* lay eggs endophytically in submerged leaves while *Ictinogomplus rapax* lays on tightly coiled filament at the pole and prevents the eggs swepting down stream (Corbet, 1980). Similarly, *Burmagomphus sivalinkensis* deposit eggs on the substance covered with gelationous material (Kumar, 1978).

According to Kakkassery (2004) the naiad of dragonfly pass through 10-20 stages. The naiad feed on small insects, fishes, tadpoles and redworms etc. Prereproductive adult stage lasted for few days to 2-3 months in some dragonflies and colour change to body has taken place within same sex or between sex and males became more colourful. In the present form also colour change taken place within the unmatured and matured sex and between sex. Males were uniformly pink coloured and females with blurish spots and light yellowish colouration. It can easily complete its life cycle in domestic water storage and consume mosquito larvae. Hence, this is good biocontrol agent of mosquitoes, *Aedes eagypti.*

Conclusion

1. *C.s. servilia* is rearable in domestic water.
2. It has good potential to control *Aedes eagypti* mosquitoes.
3. Need its utilization in Biological control of insect pests of economic importance.

vi) Adult Emergence Behaviour in *Crocothemis servilia servilia* Drury

For adult emergence behavioral studies, 25 well developed naiads have been observed. Well developed naiads stopped feeding and started searching a suitable site or substratum for further development. They crawled in upward direction, come out the water body, stop the rectal respiration and start spiracle respiration, select a suitable site for moulting, tightly hold the substratum by using their legs, wait for a few minutes and slowly break the weak point of mesothorasic segment dorsally, by that internal part of cuticle comes out. Head region pull back with folded wings. The complete body of adult came out (Figure 39) from mesothorasic segment. Newly emerged adult was faint coloured (Figure 40) *i.e.* whitish, both pairs of wings were attached to each other, and became dry, dried wings were slowly separated from each other and process of development from nymph to adult was completed. Newly emerged dragonfly hold its own cast between the legs up to the first light in sun. All developmental processes were completed during early morning at 3 a.m. to 6 a.m., occasionally emergence was completed in day time also in morning hours.

It can fly out within 1 hour after emergence as an adult. Emergence period averaged 2.5 hour.

Behaviours are intricate and essential for survival of dragonflies (Corbet, 1999). Metamorphosis from the nymph to adult is dramatic life change towards both morphology and behaviour (Corbet, 1980). According to Corbet (1999) the embryo develops inside the egg, several blocks of embryonic tissues stop developing and remained in a state of limbo until the nymph molts into its pre-adult stage. These tissues are called imaginal discs which produced structures that were only functional in adult stage. Rudimentary adult organ developed during the last pre-adult stage of growth when they remained under the nymphal skin and deep

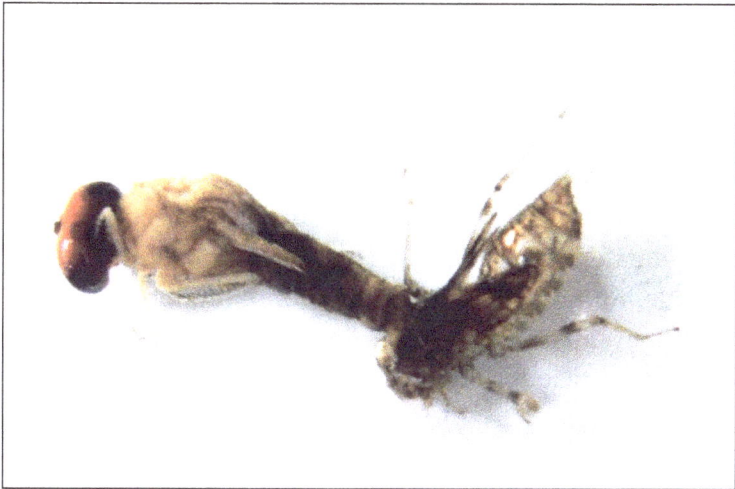

Figure 39: Adult emerging from last instar naiad in *C.s. servilia*.

Figure 40: Newly emerged adult in *C.s. servilia*.

within the body. According to Sathe and Shinde (2008) before adult emergence, most of processes are completed by the dragonfly, namely immature gonads are formed, eye buds developed into larger adult eyes, the discs that produce secondary sex organs specially ducts and copulatory structure grew in size but do not differentiate. Highly branched tubes developed throughout the body to become network of respiratory tracheae for breathing air when the naiad leave the water; when all is ready, glands in thorax produced a hormone that triggers rapid growth and differentiate imaginal discs. The brain stopped producing juvenile hormone, glands in the thorax produced another hormone causing nymphal dermis to secrete enzyme digesting cuticle for facilitating adult emergence.

The nymph swallowed large amount of water before it crawled out of aquatic habitat, it may crawl anywhere 10-15 cm to 240-270 cm on the substrate before metamorphosis. The larva pulls in its abdomen to compress it when metamorphosis begins. This swells the thoracic region and ruptured cuticle. The incipient adult continued to compress its abdomen and began to twitch and push against inside of its old skin, until it pulled its back out. Gradually the adult emerges from the skin, pulling each legs, wing and body structure into the air. When it became free from its exuviae the dragonfly stayed relatively motionless on or near its exuvia. It forced water in its appendages for elongation and stretching legs. The teneral adult forcibly ejected excess water from its digestive tract and pumped air into the appendages to burden them. Finally, the pale teneral adult became ready to fly. Almost all events are noted in *C. servilia servilia* as earlier noted by Corbet (1999).

vii) Rearing of Paramecium as a Food for Iˢᵗ Instar Naiads of Dragonflies

Paramecium caudatum (Protozoa : Paramecidae) is very good source of food for 1ˢᵗ instar naiad of the dragonfly

Crocothemis servilia servilia (Odonata : Libellulidae) and *C.s. servilia* is very good biocontrol agent of mosquitoes. Therefore, for providing continuous food for above biocontrol agent, mass culture technique of paramecium was developed. *P. caudatum* is found in freshwater ponds, pools, ditches, steams, rivers, lakes and reservoirs. It is also abundant in stagnant water containing decaying organic matter. Paramecium feed on bacteria and some small protozoans. Mosquitoes are vectors for several draded and uncurable diseases like Malaria, Filaria, Dengue, JE, Chikungunya etc. Their control became a chronic problem since they have developed resistance against several pesticides. Hence, hoping their control by dragonflies as eco-friendly control the work was undertaken. Review of literature indicates that Giese (1945), Sonneborn (1950), Karakashian (1963), Hiwatash (1968), Fok and Allen (1979) etc. contributed on culture of paramecium.

A simple method for paramecium culture was developed under laboratory conditions (27±1°C, 75±1 per cent R.H., 10 hr photoperiod) the description of which is given in the text. Similarly, mass culture method for *P. caudatum* was also developed with the help of glass aquarium and the growth medium. Distilled water 5 lit. 500 g of folks and corn husk equal proportion, 50 drops of skimmed milk and Parle-G biscuit.

The balanced medium for growth of paramecia should contain 6 per cent solution of NaCl, KCl, $CaCl_2$, $MgCl_2$ and $MgSO_4$ in the proportion of 100, 3.2, 1.0., 7.8 and 3.8 respectively (Giese, 1945). It may be diluted for five hundred times. Sea water may be diluted two hundred and fifty times and pond water need not dilution. According to Giese (1945) some times ordinary distilled water is toxic and must be redistilled for removing copper parts and toxic volatile matters. He used hay, hay wheat, mushroom and other infusions for mass culture of paramecia in addition lettuce infusion with very good results. The present method developed is more economical from the view point of cost benefit ratio. Giese (1945) developed simple culture method for determination of division rate in 3 paramecia *viz.*, *Paramecium caudatum, P. aurclia,* and *P. multimicronudeatum.* He collected wild culture to begin with the experiment and washed paramecia for removing bacteria by the method proposed by parpart (1928). The paramecia are sterlised by sterile balanced salt solution and determined the division rate in the above species.

Kotpal (1980) suggested 1 litre of pond water with dead or decaying vegetation in a glass container for 2-3 days for obtaining paramecia. Paramecia can easily cultured through chalkey's medium containing NaCl - 80 mg, $NaHCO_3$ - 4 mg, KCl - 4 mg, $CaCl_2$ - 4 mg and CaH_4 (PO_4) H_2O - 1.6 mg, 1 litre distilled water and 7-12 drops skimmed milk.

Symbiotic associations are widely distributed throughout the plant and animal kingdom (Karakashian, 1963) but relatively little is known about the detailed nature of the interactions between symbiotic partners. The symbiotic complexes frequently possess properties, not characteristic of either partner alone are interesting to investigate for enhancing the production of individuals. A single paramecium normally harbors several hundred algae that are transmitted to both daughter cells at division and retained during conjugation. Investigating metabolism of paramecium in such conditions would help mass culture in future.

Souneborn (1950) prepared exhaustive culture medium for paramecia. He collected paramecia by centrifugation. Recently, Fok and Allen (2007) contributed on mass culture and structurve of axenic *Paremecium caudatum*. For establishing and growing *P. caudatum* in mass axenic culture, the culture medium of Soldo, Godoy and van wagtendonk was modified by these workers by substituting phosphatidylethenolamine (PE) for TEM-4T and by a 10 fold increase in folic acid. They obtained population densities of 4000 to 6000 cells/ml and a generation time of 20-26h regularly. Optimal growth was obtained with PE-Stigmasterol ratios between 40:1 to 400:1. *C.s. servilia* and *Bradinopyga geminata* (Rambur) are promising predators of mosquito larvae at 3rd, 4th and at later instars. However, Ist instar naiads of dragonfly species do not feed on mosquito larvae but need paramecian food. Therefore, the present mass culture method of paramecium will add great relevance to support the concept of biological control of mosquitoes through dragonflies.

viii) Rearing of Daphnia as a Food for IInd and IIIrd Instar Naiads of dragonflies

Daphnia carinata is very good source of food for second instar naiads of dragonflies *Crocothemis servilia servilia* Drury and *Bradinopyga geminata* Rambur (Odonata : Libellulidae). The above two species of dragonflies are important biocontrol agents of mosquitoes. Control of mosquitoes became chronic problem due to pesticidal resistance in mosquitoes. Nutritional requirement is prime requisite for mass rearing of bicontrol agents and in any biocontrol programme of pests mass rearing of biocontrol agent is soul component. Mosquitoes spread many draded diseases such as Filaria, Malaria, Dengue, JE, Chikungunya, etc. Hence, the present work was objected primarily to develop mass culture of Daphnia and secondly to develop mass rearing technique of dragonflies for control of mosquitoes. Perusal of literature indicates that Smith (1963), Alan *et al.* (1968), Dinges (1973), Irgolic *et al.* (1977), Pauw *et al.* (1981), Lynch (1989), Mokry and Hoagland (1990), Elendt and Bias (1990) etc. attempted the work related to biology and mass culture of Daphnia.

A simple mass culture method for Daphnia has been developed under laboratory condition (27±1°C, 75±1 per cent R.H., 12 hr photoperiod) the description of which is given under materials and methods. Per litre of water about 10,000 Daphnia were produced and in 5 litre of water taken in glass aquarium produced about 50,000 Daphnia within 6 weeks. Green algae *Selenastrum* sp. of dry weight 1.15 mg per litre test water yielded highest number of Daphnia, 12,000 within 6 weeks.

Hardy and Duncan (1994) studied the effects of food concentration and temperature on embryonic and post embryonic duration of three tropical species, *Daphnia gessneri, Diaphanosoma sarsi* and *Moina reticulata*. They reported that in both embryonic and postempbryonic durations there was a body size effect as the absolute durations were longest in the largest species and shortest in smallest species.

Goulden *et al.* (1982) given procedures and recommendations for the culture and use of Daphnia in bioassay studies. They reported that growth, reproduction and survival in Daphnids are very dependant upon the quality and quantity of food. They recommended algal as well as artificial diet for Daphnia culture. They used three types of food *viz.* i) 1:1 mixture of two algal species *Ankistrodesmus falcatus* and *Chlamydomonas reinhardtii* (ii) *Selenastrum capicornutum,* a green alga. (iii) Trout chow (PR-II) and baker's yeast, an artificial food diet. They recommended the diet of green algae for culture of *Daphnia.* 1.25 mg dry weight of algae per lit of test water found effective for culture of *Daphnia.* They further recommended that life cycle tests be run for 21 rather than 28 days. Nebekar (1982) evaluated life cycle test method with silver and endosulfan. He reported that feeding rates of 30 mg food 1^{-1} test water depleted dissolved oxygen and vitamin enriched algae facilitates for good culture of Daphnia.

Mount and Norbeg (1984) studied a seven day life cycle Cladoceran toxicity test. They says that *Ceridaphnia reticulata* can be easily cultured; its handling, control and survival is not the problem. Paul and Crease (1983) studied clonal diversity in populations of *Daphnia pulex* reproducing by obligate parthenogenesis. They identified 39 clones in 21 habitats and found marked Hardy - Weinberg deviations, gamatic imbalance and high hetrozygosity. Irgolic *et al.* (1977) cultured *Daphnia magna* in a medium containing 74 As-labeled H_3AsO_4 and 1 ppm Na_2HASO_4 expressed as arsenic. They studied characterization of arsenic compounds formed by *Daphnia magna* and marine alga *Tetragelmis chuii* from inorganic arsenate. The arsenic metabolites were extracted with a chloroform - methanol solution and isolated by using TLC, revealed the presence of a 74As - containing product which migrated with phosphati-dylethanolamine. Dinges (1973) studied the ecology of Daphnia in stabilization ponds. He observed that the presence of Daphnia resulted in clearified wastewater effluent. Dinge (1973) studied the ecology of two Daphnia species namely *D. similis* and *D. pulex* which were only present in extra deep ponds. *Daphnia* sp. serves to stabilize and clearify water by digestion, incorporation of organics into biomass rendering material inert by conversion into chitins. Hence, they are feasible for clearing ponds.

Mokry and Hoagland (1990) studied acute toxicities of five synthetic pyrethroid insecticides to *D. magna* and *Ceriodaphnia dubia*. They reported that four generation pyrethroids exhibited greater toxicity towards Daphnia. Therefore, pyrethroid toxicity be avoided during mass culture. In the present study mass culture is made at laboratory conditions and with no contamination of pyrethroid or other insecticides. According to Pauw *et al.* (1981) *D. magna* can be successfully grown on rice bran. No deficiencies were noted after more than 20 generations. The conversion ratio from rice bran to *Daphnia* biomass (wet weight) was found to be between 2 and 1 to 1, depending on the cultivation method used. Energy content of rice bran grown Daphnids was 17000 $J/9^{-1}$ dry weight and protein content was 45 to 50 per cent of dry material.

Parasitism to Daphnia is limiting factor for mass culture. Therefore, Ebert *et al.* (1996) studied phylogenetic position of *Pasteuria ramose* Metch., a parasite of *Daphnia magna* Straus and placed under Bacteria. The parasite was confused with protozans, yeast and Bacteria. Parasitism can be avoided under laboratory condition by taking clear water. Elendt and Bias (1990) studied trace nutrient deficiency in *D. magna* caused by selenium. They concluded that ion-exchanger cartridges can leach detrimental substances into the matrix water. Therefore, only ultrapure matrix waters should be used for preparation of culture media. Reeve (1963) studied filter feeding in *Daphnia*. *Daphnia* is capable of regulating its rate of feeding in a such way that, as the plankton density increases, the filtration rate maintains a constant maximum value while the ingestioin rate increases, and when the density reaches a value at which a constant maximum ingestion rate is attained, the filtration rate falls off. The aspects discussed earlier clearly indicate that Daphnia is naturally adaptive and easily rearable and usable as food for biocontrol agents of mosquitoes like dragonfly naiads.

ix) Rearing of Mosquitoes as a Food for IIIrd Instar onward Naiads of Dragonflies

Mosquitoes cause huge medical and financial loss by spreading diseases like Yellow fever, Filaria, Malaria, Japanese Encephalitis, Dengue, Chikunguniya, etc. The mosquito *Aedes aegypti* (Linn.) is very potential vector of Dengue and Chikungunya in India. Out of 3500 species recorded from the world only about 200 species bite or bother humans (Fang, 2010). In recent days Dengue and Chikungunya became an epidemic burden for Western Maharashtra and Marathwada region of Maharashtra. For biological control of mosquitoes required large number of mosquitoes for natural enemies as food to them. The Dragonflies *Bradinopyga jaminata*, *Pantala flaviscens* and *Crocothemis servilia servilia* are good natural predators of mosquitoes. Specially 3rd instar onwards the dragonfly naiads consume large amount of mosquitoes in aquatic habitat. Therefore, for mass rearing of dragonflies, mass rearing of mosquitoes is essential. Hoping control of mosquitoes by biological means the present work was carried out. The review of literature indicates that the rearing of mosquitoes has been attempted by Igbinosa (1989), Spitzenf Takken (2005), Craig (1964), Fang (2010), Clements (1992), Akov (1964), Borman (1967), Brust and Kalpage (1967), Dimond *et al.* (1956), Fay (1964), Granett and Haynes (1944), Gerberg *et al.* (1968, 1969), Johnson (1937), Rozeboom (1936), Wallis and Speilman (1953) etc.

Screening of mosquitoes has been made in the laboratory for which windows were always locked and screened with nylon mosquito net. The used lab was kept empty from other materials and the room was painted with white light colour specially, walls and ceiling. For rearing larvae and pupae declorinated and deoxygenated water was used. For mating 3 days old separately maintained 5 males were introduced in a container of one litre size containing separately reared 20 females. The males were fed with 100 per cent honey and females with water with cotton ball and honey sponge respectively. After mating females were allowed to lay eggs on moist filter paper kept on 1 lit. beaker. The eggs were kept in glass cage (25 × 25 × 25 cm) for 48 hours. The beaker/container was removed after 48 hours and then egg paper was kept for 24 hours and then dried for 4 days at 27 ±1°C and 75-80 per cent R.H. Eggs can be stored for longer period, months together by maintaining temperature and humidity. The eggs hatched better in deoxygenated water within 5 to 60 minutes. For deoxygenation of water add the Brewer's yeast to the chlorinated water and leave it overnight with container tightly closed. After deoxynation of water a filter paper with eggs be totally submerged keeping eggs on the upper side of the paper. Eggs were hatched within 20-30 minutes. Unhatched eggs are allowed to dry, and submerge once again for hatching eggs. After 2 hours the hatched larvae were transferred into the aquarium with dechlominated water (aged). Fill the aquarium with dechlorinated water and a little food to the 1st instar mosquito larvae. Larvae were counted by the aliquot method and 60-70 larvae/sq.cm of water surface with a depth of 1.5 cm have been released into the aquarium. Thus, one aquarium of size 3 × 2 × 1 ft with litre of water reared larvae of mosquitoes.

There were 4 instars in the mosquito. First, second, third and fourth instars lasted for 2-3 days, 2-3 days, 1-2 days, 2-3 days respectively. 7 to 8 day was larval period. Pupal period was 2-3 days. Newly emerged adults were fed with 50 per cent sugar solution soaked in cotton balls. Surface scum was removed 12 hours interval. Aged water (dechlorinated) was added every day in aquarium with food to mosquito larvae. The larval food refers to combination of following:

☆ 25 mg of powdered dry brewers yeast

☆ 25 mg of Bacto-brain heart.

For first instars, a food containing powdered yeast and powdered dog food in 1:1 proportion was given. The proportion of food in days is given below.

0 - 1 day old	0.2 mg/larva
2 day old	0.3 mg/larva
3 day old	0.4 mg/larva
4 day old	0.6 mg/larva
5 day old	0.6 mg/larva
7 day old	0.6 mg/larva

On the 6ᵗʰ day larvae were separated for separation of pupae for further experiments. Pupae were given no food as they can not take.

The pupae were counted and placed in plastic container and or aquarium (size 3 × 2 × 1 ft.). Approximately 50,000 pupae were placed in above aquarium and 1000 pupae in an aquarium of size 38 × 23 × 23 cm (Figure 41) by maintaining sex ratio (m:f) 1 : 3, 1 pupa/2 sq.cm. Adults were given 50 per cent honey and blood meal of rat per day for 2-3 hours. The adults survived 22 days. During the rearing temperature was set to 27 ±1°C, 75-80 per cent R.H. and 12 hr photoperiod.

Glass aquariums (Figure 42-45) size of 45 × 30 × 30 cm of two partitions (each of size 22 × 30 × 30 cm) having circular hole of diameter 18 cm for adult mosquito entrance

Figure 41: Mosquito rearing aquarium.

was used for providing blood meal of rat to mosquitoes in the laboratory. In one compartment rearing of last instar mosquito larvae were allowed to continue for adult emergence and sufficient adults were collected and in another compartment one rat was kept for 2-3 hours/day for mosquito blood meal. Mosquitoes entered in to the compartment of rat through circular hole and enjoyed rat blood meal easily. After 2-3 hours of exposure, the rat was removed and wet filter paper was placed in to the compartment of rat and mosquitoes were allowed to lay eggs on the wet filter paper. Thus, sufficient eggs were collected on the filter paper. The eggs hatched within 12 hours and newly emerged mosquito larvae either submerged in the same aquarium or in the other compartment of the aquarium or placed for rearing in other aquariums of size 46 × 30 × 30 cm or 38 × 23 × 23 cm or 3 compartmental aquarium of size 45 × 30 × 30 cm (each 15 × 30 × 30 cm, in

Figure 42: Mosquito oviposition unit with rat.

Dragonflies Production Technology

Figure 43: Mosquito rearing cages with oviposition unit and rat.

Figure 44: Three compartmental glass aquariums for mosquito rearing.

length, width and height respectively). Thus, sufficient number of mosquito larvae were obtained in aquariums and provided to naiads of dragonflies.

Morlan (1966) suggested that hatching of eggs is better when they immersed in a 24 hour old mixture of 0.1 g of brewer's yeast per litre of water. For blood meal Morlan

Figure 45: Different rearing aquariums for mosquito larvae rearing.

(1966) used a rabbit for 2-3 hours each day for mass rearing. In the present study mouse was used for 4 hours exposure to adult mosquitoes for blood meal and oviposition. Entomologists often require large number of experimental mosquitoes for laboratory studies. The quality of experimental mosquitoes should be high and constant, otherwise the result of replicate experiment be contradictive or results from subsequent field work might be disappointing. Any minor change in technique have drastic results both, positive and negative. The quality of mosquitoes influenced by population density, nutrition, environmental conditions etc. Therefore, appropriate conditions and correct technology is needed. The rearing of mosquitoes is complex and demanding for several reasons. Mating was not necessary accomplished naturally and females need a blood meal to develop eggs - in malarial mosquitoes. Because of tropical conditions mosquito larvae developed fast and need to cared for daily. Therefore, climatic chambers of mosquitoes were kept warm and sweaty. The laboratory of Wageningen has cultured different colonies of malarial vector mosquitoes successfully for many years for which rearing conditions were 27°C temperature, 80 per cent relative humidity and light and dark regimen were shifted from 06.00 - 18.00 towards 00 : 20 - 12.00. In the present study rearing conditions were 27±1°C, 75-80 per cent R.H. and 12 hr photoperiod. *Anopheles stephensi* was reared under insectary conditions, 27°C, 75 per cent R.H. and 12 L. 12 D photoperiod by Gerberg *et al.* (1968). Rozeboom (1936) reared *Anopheles albimanus* Wied. at 27°C by providing dechlorinated water, powdered Durina Laboratory Chow supplemented with Fleischmann's yeast. The larval period at such condition he noted was 7-14 days. He provided a guinea pig as blood meal and honey saturated with cotton wool balls. The adult longevity was 30 days. In the present study adults were provided rat for blood meal 2-3 hours for each day and 50 per cent honey and adults survived for 22 days. While, in *Culex salinarius* Coquillett Wallis and Spielman (1953) provided fresh apple slices daily and blood meal preferably a bared foreavm. In *C. salinarius* eggs were deposited 18 days after adult emergence while in present form within 3 days only. The present mass rearing technique is useful for mass rearing of dragonflies *Bradinopyga jaminata* for biological control of mosquitoes.

x) Mosquito Larvae Consumption Rate by a Dragonfly *Pantala flaviscens*

Dragonflies (Order : Odonata) are referred as mosquito hawks due to their predaceous behaviour. Adult dragonfly consume adult mosquitoes while naiads of dragonfly consume mosquito larvae in aquatic habitats. However, according to Bay (1974) dragonflies do not consume enough mosquitoes to cause a significant impact on mosquito populations in nature. Therefore, it will be very interesting to find out the consumption rate of mosquitoes by the dragonfly naiads *Pantala flaviscens* for developing rearing technique for *P. flaviscens*.

Review of literature indicates that biodiversity of mosquitoes from Kolhapur district has been studied by Sathe and Girhe (2001) and Sathe and Tingare (2010) and the studies related to mosquito consumption by dragonflies have been attempted by Laird (1973), Bay (1974), ElRayah (1975), Santamarina and Mijares (1986), Urabe *et al.* (1986), Thomus *et al.* (1988), Sebastian *et al.* (1990), Corbet (1999), Sathe (2006, 2008) etc.

The results recorded in Table 4 indicates that 3rd, 4th, 5th, 6th, 7th, 8th, 9th and 10th instar, naiads consumed 11.89, 14.00, 18.20, 20.40, 22.60, 24.80, 27.20 and 30.40 larvae of mosquitoes within 24 hours. The rate of consumption was found in increasing order with respect to older instars. In 6th and 7th instar no much difference in consumption rate was observed. However, the last instar larvae have consumed greatest number of larvae. Naiads of 1st and 2nd instar have not responded to the mosquito larvae after starvation of 24 hours. Mosquito larvae comsumption in *C. sevilia* and *B.jaminata* are recorded in Table 5.

Table 4: Mosquito larvae consumption by *P. flaviscens*.

Naiad Instar	No. of Mosquito Larvae given	No. of Mosquito Larvae Remain (unconsumed) after 24 hrs.	Per cent of Mosquito Larvae Consumed within 24 hrs.
3rd	500	441	11.8 per cent
4th	500	430	14.0 per cent
5th	500	409	18.2 per cent
6th	500	398	20.4 per cent
7th	500	387	22.6 per cent
8th	500	376	24.8 per cent
9th	500	364	27.2 per cent
10th	500	348	30.4 per cent

* Replicates five, each containing 100 larvae.

Table 5: Mosquito larvae consumption by *C. servilia servilia* and *B. jaminata*.

Naiad Instar	No. of Mosquito Larvae	No. of Mosquito Larvae Consumed after 24 hrs.		Per cent of Mosquito Larvae Consumed within 24 hrs.	
		C. servilia servilia	*B. jaminata*	*C. servilia servilia*	*B. jaminata*
3rd	500	50	52	10.00	10.4
4th	500	65	68	13.00	13.6
5th	500	76	80	15.2	16.00
6th	500	98	102	19.8	20.4
7th	500	129	138	25.80	27.60
8th	500	142	151	28.4	30.20
9th	500	165	163	33.00	32.60
10th	500	164	164	32.80	32.80

* Replicates five, each containing 100 larvae.

There are two approaches to the biological control of pests namely, inoculation and inductive or augmentative release (AR) (Sathe and Bhoje, 2000). Both approaches are useful for mosquito control. The success of above approaches are dependant on consumption rate of preys by predator (biocontrol agent). Dragonflies (Odonata) are predators of several insect pests. Houseflies, paddy moths, jassids and mosquitoes are favourite preys of dragonflies (Metcalf and Flint, 1979; Sathe, 2006, 2008). Therefore, Thomas *et al.* (1988) studied predatory efficiency of nymphs of *Bradynopyga jaminata* and *Brachythemis contaminata* on mosquito larvae. They reported that both above mentioned species of dragonflies have good potential of mosquito larvae consumption and they concluded that above two species could be used in biological control of mosquitoes specially, *Aedes* spp.

In Cuba, Santhamarina and Mijares (1986) made field and laboratory observations to study the efficacy of nymphs of *Pantala flaviscens* and *Tramea abdominalis* against the larvae of mosquito *Culex quinquefaciatus*. Their preliminary studies suggested good potential for mosquito larvae consumption by above species of dragonflies.

Urabe *et al.* (1986) studied and evaluated the predatory capacity and efficiency of *Sympetrum frequens* against the larvae of *Anopheles sinensis* in laboratory and reported that when the nymph size increased, subsequently the number of mosquito larvae consumption also intensified. In the present study same situation was noticed. Urabe *et al.* (1986) also studied predator prey relationship between *S. frequens* and *An. sinensis* in rice field near Omiya, Japan has revealed that the density of mosquito larvae increased when the nymphal density of the predator become low. The distribution pattern of the predators and prey in the smaller field had a non-overlapping tendency, indicating effective predation. Similarly, Sebastian *et al.* (1990) demonstrated the trials of consumption of *Aedes aegypti* mosquito larvae by the naiads of dragonfly *Czecontenemis servelia* in Rangoon, Myanmar. More than 92 per cent of the local population of *A. aegypti* was controlled by this operation. The present predatory species is potential feeder of mosquito larvae and usable for controlling *Aedes* mosquitoes and their after the mosquito borne diseases.

xi) Rearing of Dragonflies

Rearing of Dragonflies in Large Glass Aquariums (Figure 46)

Matured naiads of *Crocothemis servilia servilia* Drury have been collected from Shivaji University campus (tanks) (Figure 15) and reared in glass aquarium (size 3 ft × 2 ft × 1 ft.) at laboratory conditions (27±1°C, 80 per cent R.H. and 14 hr photoperiod) individually. Newly emerged adults were kept individually in 1 litre plastic container for some time and then one male and one female released in the large insect net 25' × 15 ' × 10' (in Length, width and height) fitted on the rearing water tank of size 20 × 10 × 2.5 ft. (in L, W, H) for mating purpose for 24 hr. Then female has been isolated and allowed to lay eggs in water in petridish by deeping abdomen. Sufficient number of eggs have been collected on wet filter paper. Eggs hatched in about 5-8 days. The first instar naiads are very small sized and light coloured. They are taken in 1 lit plastic container for rearing purpose. For first instars paramecium is best food. Paramecium can be cultured as per the method given earlier.

Figure 46: Rearing of dragonflies in glass aquariums.

The plastic container should contain little duck weed/ Hydrilla for aeration. Within 2-3 days first instars moult into second stage (instars). The second instars are now released into a glass aquarium of size $45 \times 22 \times 48$ cm (L,W,H) and provided Daphnia as food in addition to some Hydrilla plants for aeration. This stage lasts for 3-5 days. Then third instar onward naiads are transformed to large glass aquarium (size $3 \times 2 \times 1$ ft (in L, W, H). 3rd to 6th instars are given mosquito larvae for feeding purpose. They are provided either mosquito larvae/Mud worms/grubs of bees as food. There are 12 instars in *C. servilia servilia*. For last instar, wooden reeds are fixed in the glass aquarium so that the naiads can climb upon them for fixiing body and casting exuviae and for adult emergence. Adult emergence process was completed within 2 to 3 hours mostly from, 3.00 a.m. to 6.00 a.m. or sometimes at morning hour. Sufficient number of naiads and adult dragonflies can be had by this method.

Rearing of Dragonflies in Water Tank Covered Net

Newly emerged pair (male and female) of *Crocothemis* were released in the water tank covered with net size, (25 × 20 × 10 ft) for mating for 24 hr. Later, 5 to 10 mated females were released in a cage containing water tank (size 20 × 15 × 2.5 ft. (in L,W,H) for 24 hr for egg laying. The females will start egg laying in the tank water. Thus, sufficient number of eggs will be received for further rearing of naiads. The tank was provided with some plants of duck weed/Hydrilla for aeration for naiads. Paramecium and Daphnia cultures were made available at the bottom of tank, for first two instars and *Aedes* mosquito larvae culture for 3rd -6th instars of dragonflies. Then after, mud worms/bee grubs/red worms/mosquito larvae were given to instars 7 to 12 as per the need mostly mosquito larvae. Final instar naiads will climb on the walls of tank for attachment of their exuviae and for facilitating adult emergence. Within 2-3 hr adult emergence process was completed and dragonfly was ready to fly after spreading the wings and stretching legs and body at every morning in the breeding cages. Thus, large number of naiads in the tank water and adults in the large insect cages were reared.

Rearing of *Bradinopyga jaminata* and *C. s. servilia*

The basic requirement for both the species was same except the duration of life cycle of the species.

		B. jaminata	*C. s. servilia*
Eggs laying	-	200-250	180–200
Incubation period	-	5-8 days	10–12 days
Larval period	-	45-54 days	47-56 days
Total instars	-	10	12

		B. jaminata	C. s. serviia
1st instar	-	2–3	2–3
2nd instar	-	3–5	3–4
3rd instar	-	5–6	5–6
4th instar	-	5–7	5–7
5th instar	-	5–7	5–7
6th instar	-	5–7	5–7
7th instar	-	5–7	5–6
8th instar	-	4–6	5–6
9th instar	-	4–5	3–5
10th instar	-	2–3	3–5
11th instar	-	0–0	3–5
12th instar	-	0–0	2–3
Progeny production/female (Fecundity)		188	176

xii) Mass Rearing Technique of Dragonflies

Two dragonfly breeding water tanks have been constructed in open environment (field) in Shivaji University, Kolhapur Campus, behind Department of Zoology, Shivaji University, Kolhapur. The size of first tank was 20 × 10 × 2.5 feet (Figure 47) and another was 10 × 10 × 2 feet (Figure 48) in length, width, and height respectively, both tanks were covered by a nylon bamboo net/cage of size 25 × 15 × 10 and 15 × 15 × 10 length, width and height respectively. The cage was made quadrangular as green house. The cages were facilitated by entry doors for man for taking observations. The tank was constructed by bricks with 10 inch wall width and plastered from bottom to all the sides, internally and externally. A pair (one male and female) of dragonfly species was introduced in to each nylon bamboo net/cage (Figure 49) for mating and oviposition purpose. Before introduction of both the sexes in a cage they were kept isolated in glass cage (Figure 20) individually for 24 hour providing adult mosquitoes *Aedes aegypti* as food. After introduction in the cage the sexes mate easily and a female started egg laying in the tank water. Within 5 to 6 days eggs hatched in to first instar naiads. They started feeding on paramecium introduced 5,00,000/day at the bottom of tank. Second and third instars were fed on daphnia introduced in the tank 2,00,000/day. Later instars offered first to fourth instar larvae of mosquitoes which were reared outside and inside the tank. Mosquito larvae have been introduced daily 3000-5000 in to the tank. The adult mosquitoes emerged were consumed by dragonfly adult in the nylon cages. On an average a single female of *B. jaminata* and *C. sevilia sevilia* (Figure 50) produced 180 and 170 adult individuals and 100 females easily produced on an average 17000 to 18000 dragonflies within couple of months. The sex ratio was found favoring males.

Precautions during Rearing of Dragonflies

1. Frog and spider should be prevented from their entry into the water tank and nylon cages covered over the tanks respectively,

2. Timely and instar wise specific food be given to the naiads of dragonflies for healthy growth and avoiding cannibalism.

3. In single tank single species and single generation bee reared for avoiding cannibalism.

4. 75 per cent tank water be replaced per week with the help of rubber pipe through cotton clothing filtration.

Figure 47: Water tank with nylon cage.

Figure 48: Water tank with nylon cage of size 10 × 10 × 2 feet.

Figure 49: Male female on water tank in rearing cage.

Figure 50: Dragonflies in nylon net cage.

POT/SCREEN/GLASSHOUSE/ FILED EXPERIMENTS

Field Breeding Tanks for Dragonflies

Two dragonfly breeding water tanks have been constructed in open environment (field) in Shivaji University, Kolhapur Campus, behind Department of Zoology, Shivaji University, Kolhapur. The size of first tank was 20 × 10 × 2.5 feet and another was 10 × 10 × 2 feet in length, width, and height respectively, both tanks were covered by a nylon bamboo net/cage of size 25 × 15 × 10 and 15 × 15 × 10 length, width and height respectively. The cage was made quadrangular as green house. The cages were facilitated by entry doors for man for taking observations. The tank was constructed by bricks with 10 inch wall width and plastered from bottom to all the sides, internally and externally. A pair (one male and female) of dragonfly species was introduced in to each nylon bamboo net/cage for mating and oviposition purpose. Before introduction of both the sexes in a cage they were kept isolated in glass cage individually for 24 hr. providing adult mosquitoes *Aedes aegypti* as food. After introduction in the cage the sexes mate easily and a female started egg laying in the tank water. Within 5 to 6 days eggs hatched in to first instar naiads. They started feeding on paramecium introduced 5,00,000/day at the bottom of tank. Second and third instars were fed on daphnia introduced in the tank 2,00,000/day. Later instars offered first to fourth instar larvae of mosquitoes which were reared outside and inside the tank. Mosquito larvae have been introduced daily 3000-5000 in to the tank. The adult mosquitoes emerged were consumed by dragonfly adult in the nylon cages. On an average a single female of *B. jaminata* and *C. servilia servilia* produced 180 and 170 adult individuals and 100 females easily produced on an average 17000 to 18000 dragonflies within couple of months. The sex ratio was found favoring males.

Precautions during Rearing of Dragonflies

1. Frog and spider should be prevented from their entry into the water tank and nylon cages covered over the tanks respectively,

2. Timely and instar wise specific food be given to the naiads of dragonflies for healthy growth and avoiding cannibalism.

3. In single tank single species and single generation bee reared for avoiding cannibalism.

4. 75 per cent tank water be replaced per week with the help of rubber pipe through cotton clothing filtration.

Frontline Field Demonstration

1. Area coverage in hectares: not applicable.

2. Corps covered : not applicable

3. Biocontrol agents/biopestiscieds developed and mass production technique of them developed in the project.

   Mass production technique have been developed for paramecium (*Paramecium cudatum*), Daphnia (*Daphnia carinata*), Mosquitoes *Aedes aegypti* and dragonflies , *Bradinopyga jaminta* and *Crocothemis servilia servilia*.

iv) Details of filed demonstration of the biocontrol agents:

The present project is restricted to mass rearing of dragonflies. Hence, the commercialization and utility of these biocontrol agents will be the extension part of the project in future.

Conclusion

Chemical control never solve the problem of pests permanently but biocontrol agents can do. Secondly, pesticides lead several serious problems like pollution, health hazards, killing of beneficial organisms, pest resistance, secondary pest outbreak, pest resurgence, interruption to ecocycles. Hence, Biocontrol agents are living weapons over chemical control and keep the environment healthy and ecofriendly. Therefore, biocontrol agents like dragonflies be mass reared and fully utilized to control mosquitoes and various insect pests including paddy, sugarcane and mulberry crops.

Chapter 6
Publications

Papers Published in Journals during the years 2009 - 2012

1. Sathe, T.V. and A.R. Bhusnar, 2010. Biodiversity of dragonflies (Order –Odonata) from Sawantwadi region of Western Ghats, Maharastra. *Life Science Bulletin,* 7(1): 65-66. (ISSN – 0973- 5453).

2. Sathe, T.V. and A.R. Bhusnar, 2010. Biodiversity of Mosquitovorus dragonflies (Order: Odonata) from Kolhapur district including Western Ghats. *Biological Forum, an international journal,* 2 (2) : 38-41. (ISSN – 0975- 1130).

3. Sathe, T.V. and A.R. Bhusnar, 2010. Mosquito larvae consumption rate by a dragonfly *Pantala flaviscens* (Odonata : Libellulidae). *Life Science Bulletin,* 8(1) : 107-108. (ISSN – 0973- 5453).

4. Sathe, T.V. and A.R. Bhusnar, 2011. Biology of a dragonfly *Bradinopyga jaminata* (Rambur) (Order-Libellulidae), a biocontrol agent of *Aedes* Mosquitoes. *J. Adv. Zool.*, 32(1) : 9-11. (ISSN – 0253 – 7214).

5. Sathe, T.V. and A.R. Bhusnar, 2011. Adult emergence behavior in *Crocothemis servilia servilia* Drury. *Geobios*, 38, 197 – 199. (ISSN – 0251 – 1223).

6. Sathe, T.V., 2010. Biodiversity of damselflies from Koyana dam and around area. *Flora and Fauna*, 16 : 68 – 72. (ISSN – 0971- 6920).

7. Sathe T.V. 2011. Ecology of mosquitoes from Kolhapur district, India. *International Journal of Pharma and Biosciences*, 2 : (4)(B), 103-111. (ISSN – 0975 – 6299).

8. Sathe T.V. 2011. Laboratory mass culture technique for *Paramecium caudatum* (Protozoa : Paramecidae). *J. Curr. Sci.*, 16 (1) : 133-135. (ISSN- 0972 – 6101).

9. Sathe T.V. 2011. Mass culture of *Daphnia carinata* (Crustacea : Daphinidae) under Laboratory Conditions. *Indian J. Environment and Ecoplaning*, 18(2-3) : 243 – 246. (ISSN – 0972 – 1215).

10. Sathe, T.V. and A.R. Bhusnar, 2010. Rearing of dragonfly *Crocothemis servilia servilia* Drury (Odonata: Libellulidae) under control condition (Presented).

11. Sathe, T.V., 2012. Mass rearing of mosquito *Aedes aegypti* under laboratory conditions (Communicated).

Research Papers Presented in National and International Conferences during year 2009–2012

1. Sathe, T.V. and A.R. Bhusnar, 2010. Biodiversity of dragonflies (Oder-Odonata) from Amba Ghat,

Maharashtra. *National Level Seminar on Biodiversity: Emerging Challenges and Opportunities.* Majalgon Art, Science collage, Majalgon. 29-30 August 2009.

2. Sathe, T.V. and A.R. Bhusnar, 2010. Biology of a dragonfly *Crocothemis servilia servilia* Drury (Odonata: Libellulidae), a predator of paddy pests in Kolhapur. *National Conference on Life Sciences with special references to Environmental Biotechnology and Biodiversity.* Vivekanad Collage, Kolhapur.2nd – 3rd Feb. 2010.

3. Sathe, T.V. and A.R. Bhusnar, 2010. Biodiversity of Dragonflies of Western Ghats of Sawantwadi region, Maharashtra. *National Conference on Life Sciences with special references to Environmental Biotechnology and Biodiversity.* Vivekanad Collage, Kolhapur. 2nd – 3rd Feb. 2010.

4. Sathe, T.V. and A.R. Bhusnar, 2010. Biodiversity of Dragonflies (Order-Odonata) from Agroecosystems of Kolhapur region of Maharashtra. *Second International Conference* on *Bio-Wealth Management for Sustainable Livelihood.* Ranchi (Jharkhand). 20th – 22nd November 2009.

5. Sathe, T.V. and A.R. Bhusnar, 2010. Diversity of Dragonflies from plain region of Kolhapur district: *National Conference on Global Warming and Conservation Strategies,* LBS College, Satara, Dec. 21-22, 2010.

6. Sathe, T.V. and A.R. Bhusnar, 2010. Mosquito larvae consumption rate by a dragonfly *Pantala flaviscens* naiads (Order – Odonata). *National level seminar on Modern trends in Environmental Pollution and Eco-planning,* DRZ College, Aurangabad, Dec. 30-31, 2010.

7. Sathe, T.V. and A.R. Bhusnar, 2010. Adult emergence behaviour in *Crothemis servilia servilia Drury* (Odonata: Libellulidae). *National Conference on Challenges to Biodiversity Conservation.* Dayanand Science College, Latur. 5th and 6th Feb. 2011.

8. Sathe, T.V. and A.R. Bhusnar, 2010. Rearing of dragonfly *Crocothemis servilia servilia* Drury (Odonata: Libellulidae) under control condition. *National Seminar on Recent Trends and Techniques in Biosciences.* Art. Commerce and Science College, Palus. 30th Nov. – 1st Dec., 2011.

Bibliography

Alan J. Tessier, L.L. Henry, and C.E. Goulden 1963. Starvation in *Daphnia* : Energy reserve and reproductive allocation. *Limnology and Oceanography* 28 (4), 667-676.

Akov, S. 1964. The aseptic rearing of *Aedes aegypti. Bull Wld. Hlth. Org.* 31, 463-464.

Bay, E.C. 1974. Predator-prey relatioinships among aquatic insects. *Annual Review of Entomology.* 19, 441-453.

Boorman, J.P.T. 1967. Aseptic raring of *Aedes aegypti* Linn. *Nature,* 213, 197-198.

Brust, R.A. and Kalpage, K.S. 1967. A rearing method for *Aedes abserratus* (F. and Y.). *Mosq. News,* 27, 117.

Corbet, D.S. 1980. Biology of Odonata. *Ann. Rev. Entomology,* 25, 189-217.

Corbet P.S. (1999) Dragon flies: Behaviour and ecology of Odonates, Corn. Uni. Press. New and Harley Books, Great Horetesty, UK. 1-829.

Clements A.N. 1992. The biology of mosquitoes-I: Development, nutrition and reproduction. Chapman and Hall London.

Dimond J.B., Lea, A.O., Hahnert, W.E. Jr. and D.M. Delong 1956. The amino acid requirements for egg production in *Aedes aegypti*. Dinges R. 1973. Ecology of *Daphnia* in stabilization ponds. *Reports.* 69.

Ebert, D.; Rainy, P; Embley T. M. and Schoz D. 1996. Development, life cycle, ultrastructure and phylogenetic position of *Pasteuria ramose* Metch. 1888. Rediscovery of an obligate endoparasite of *Daphnia magna* Stratus. *Phil. Trans. R. Soc.* Lond, B 29, (351), 1689-1701.

Elendt B. P; and W. R. Bias, 1990. Trace nutrient deficiency in *Daphnia magna* cultured in standard medium for toxicity testing. Effects of the optimization of culture conditions on life history parameters of *D. magna.Water Research,* 24(a), 1157-1167.

El Rayah, E.A. 1975. Dragonfly nymphs as active predators of mosquito larvae. *Mosq. News.,* 35, 220-230.

Fang, J. 2010. Ecology : A world without mosquitoes. *Nature,* 466, 432-434.

Fay, R.W. 1964. The biology and bionomics of *Aedes aegypti* in the laboratory. *Mosq. News,* 24, 300-308.

Fraser, F.C. 1933, 1934 and 1936. Fauna of British India including Ceylon and Burma. Odonata, 1: XIII + 423pp.2: XXIII + 398 pp. 3: XI+ 461pp., Taylor and Francis, London.

Fraser, F.C. 1957. A classification of the Order Odonata. *R. Zool. Soc.* N.S.W. Sydney, 133pp.

6

Fok A.K. and R.D.Allen 2007. Axenic *Paramecium caudatum*. J. Mass culture and structure. *Journal of Eukaryotic microbiology*, 26(3), 463-470.

Gerberg E.J., Gentry J.W. and L.H. Diven 1968. Mass rearing of *Anopheles stephensi* Liston. *Mosq. News*, 28, 117.

Gerberg E.J., Hopkins, T.M. and Gentry J.W. 1969. Mass rearing of *Culex pipiens* L. *Mosq. News*, 29, 382-385.

Granett, P. and Haynes, H.L. 1944. Improved methods of rearing *Aedes aegypti* mosquitoes for use in repellent studies. *Proc. 31 Ann. Meeting N.J. Mosa. Exterm. Assoe.*, p. 161-168.

Giese A.C. (1945) : A simple method for division rate determination in paramecium. *Physiological Zoology*, 18(2), 158-161.

Goulden C. E; Comottor R. M; Henrickson J.A; Horing L. L. and Johnson K. L. 1982. Procedures and recommendations for the culture and use of Daphnia in bioassay studies. *ASTM Special Technical Publication*; 766, 139-160. *Can. Entomol.*, 88, 57-62.

Hardy M; and Duncan, 1994. Food concentration and temperature effects on life cycle characteristics of tropical cladocera (*Daphnia gessneri* Herbst, *Daphnosoma sarsi* Richard, *Moina reticulata* (Daday) : 1-Development time. *Acta Amazonica*, 24(1-2), 119-134.

Hiwatash K. (1968) : Determination and inheritance of mating type in *Paramecium caudatum*. Genetics.

Irgolic K. J., Woolso E.A., R.A. Stockton, R. D. Newman, N.R.Bottino, R. A. Zingaro, P.C.Kearney, R.A.Pyles, S. Maeda, W.J.Mcshane and E.R.Cox, 1977. Characterization of arsenic compounds formed by

Daphnia magna and *Tetraselmis chuii* from inorganic arsenate. *Environ. Health Perpect*; 19, 61-66.

Johnson, H.A. 1937. Note on the continuous rearing of *Aedes aegypti* in the laboratory. *Public Hlth. Repts.*, 52, 1177-1179.

Kakkassery, F.K. 2004. Dragonflies and Damselflies in Biological control. 61-77 PP, DPH Delhi.

Karakashian S.J. 1963. Growth of *Paramecium bursaria* as influenced by the presence of algal symbionts. *Physiological Zoology* (JSTOR).

Kotpal R.L. 1980. Zoology - phylum-1 : *Protozoa.*, p.186.

Kulkarni, P.P. and Prasad M. 2002. Insecta: Odonata. Zool. Survey India: Wetland Ecosystem Series No. 3: Fauna of Ujani: pp 91-104.

Kulkarni P.P., Bastawade D.B. and S.S. Talmale, 1999. Predation of dragonflies *Ictingomphus rapax*(Rambur) and *Pantala flaviscens* (Fab.) (Odonata : Anisoptera) by the giant wood spider, *Nephila maculata* (Fab.).

Kumar A. 1972. The life history of *Lestes praemorse praemorse* (Selys) (Odonata : Lestidae). *Treubia*, 28, 3-20.

Kumar A. 1972a. The life history of *Trithemis festiva* (Rambur) (Odonata : Libellulidae). *Odonatologica*, 1, 103-112.

Kumar A. 1972b. Bionomics of *Orthetrum pruinosum neglectum* (Rambur) (Odonata : Libellulidae). *Bull. Ent*, 11, 65-93.

Kumar A. 1973. Description of the last instar larvae of Odonata from the DehraDun Valley India with notes on Biology - II (sub order : Anisoptera). *Oriental Ins*, 7, 291-331.

Kumar A. 1978. Some field notes on the Odonata around a fresh water lake in Western Himalaya. *Entomon,* 2, 225-230.

Laird, M. 1973. Dragonflies versus mosquitoes again. *Mosq. News.*, 33, 462.

Lahiri, A.R. 1987. Studies on odonate fauna of Meghalaya. *Rec. Zool. Survey of India.* Misc. publ. Paper 99: 402pp.

Lynch M. 1989. The life history consequences resource depression in *Daphnia plutex. Ecology,* 70, 246-256.

Mokry L. E. and Hoagland K.D. 1990. Acute toxicities of five synthetic pyrethroid insecticides to *Daphnia magna* and *Ceriodaphnia dubia. Environ. Toxi. Chem.,* 9(8), 1045-1051.

Mount D.I. and Norberg T.J. 1984. A seven day life-cycle cladoceran toxicity test. *Environ. Toxi. and Chem.,* 3(3), 425-434.

Nebekar A. V. 1982. Evaluation of a *Daphnia magna* renewal life cycle test method with silver and endosulfan. *Water Research,* 16(5), 739-744.

Paul Herbett and Teri Crease 1983. Clonal diversity in populations of *Daphnia pulex* reproducing by obligate parthenogenesis. *Heredity,* 51, 353-369.

Pauw N. D., Laureys P., and J. Morales 1981. Mass cultivation of *Daphnia magna* Straus on rice bran. *Aquaculture,* 25(2-3). 141-152.

Prasad M. 1995. On a collection of Odonata from Goa, India, *Fraseria* (N.S.), 2 (12), 7 - 8.

Prasad M. 1996. An account of the Odonata of Maharashtra state, India. *Rec. Zool survey India*: 95 (3 – 14), 305 - 327.

Prasad M. 1999. Faunal diversity in India: Odonata. Zoological Survey of India, Calcutta, 177-178.

Prasad M. and P.P. Kulkarni 2001. Insecta: Odonata Zool. Surv. India: fauna of conservation Area series 11: fauna of Nilgiri biosphre Reserve: 73 – 83 pp.

Prasad M. and P.P. Kulkarni 2002. Insecta: Odonata. Zool. Surv. India: Fauna of Eravikulam National Park. Conservation Area series No. 13: 7-9.

Prasad M. and Varshney, R.K. 1995. A checklist of Odonata of India including data on larval studies. *Oriental Ins.,* 29, 385 – 428.

Reeve M.R. 1963. *J. Exp. Biol.,* 40, 195-205.

Rozeboom L.E. 1956. The rearing of *Anopheles albimanus* Wiedemann in the laboratory. *Am. J. Trop. Med.,* 16, 471-478.

Sahayaraj, 2004. Indian Insect Predators in Biological Control, pp. 336, DPH, New Delhi.

Santamarina, H. and Mijares, A. 1986. Odonata as bioregulators of the larval phase of mosquitoes. *Revi.cubana de. Medina-Tropical,* 38111, 89-97.

Sathe T. V. and B.E. Girhe 2001. Mosquitoes and diseases Daya Publishing House, New Delhi pp. 1-120.

Sathe T.V. and K.P. Shinde 2006. On a new species of the genus *Rhyothemis* Hagen (Order: Odonata) from India. *J. Nat. Con.* 18 (2) 421- 424.

Sathe T.V. and K.P. Shinde 2007. On a new species of the genus *Crocothemis* Brauer from Western Ghats, Maharashtra. *Flora and Fauna,* 13 (2), 367-370.

Sathe T.V. and K.P. Shinde 2008a. Dragonflies and pest management. Daya Publishing House, New Delhi. pp-1-179.

Sathe T.V. and K.P. Shinde 2008b. Biodiversity of dragonflies from Sindudurga district, Maharashtra. *Proc. Nat. Sem. Recent. Trends Life sci. Belgaum 2*, 87-88.

Sathe T.V. and Tingare B.P. 2010. Biodiversity of mosquitoes. Mangl. Publ., New Delhi, 1-217 pp.

Sathe T.V., A. T. Sathe and Mahendra Jagtap. 2011. Mosquito borne diseases. Mang. Publ. New Delhi. pp-1-350.

Sebstian, A; Myint sein, Myat Thu. And Philip S. Corbet. 1990. Suppression of *Aedes aegypti* (Diptera: Culicidae) using augumentative release of larvae Odonata: Libellulidae with community participation in Yangon, Myanmar. *Bull Ento Res.*, 80, 223-232.

Silsby J. 2001. Dragonflies of the world: Natural History Museum and Plymbridge Distritutors Ltd. 1-124.

Smith F.E. 1963. Population dynamics in *Daphnia magna* and a new model for population growth. *Ecology*, 44(4), 651-653.

Smith K.G.V. 1973 Insects and other arthropods of medical importance. T. B.M. London pp 37 – 107.

Spitzen J. and W. Takken. 2005. Malaria mosquito rearing, maintaining quality and quantity of laboratory reared insects. *Proc. Neth. Entomol. Soc.Meet.*, 16, 95-100.

Sonneborn T.M. (1950) : Methods in the general biology and genetics of *Paramecium aurelia*. *Journal of Experimental Zoology*, 113, 87-147.

Thomas. M; Daniel, M.A. and Gladsutone, M. 1988. Studies on the food preference in three species of Dragonfly naiads with particular emphasis on mosquito larvae predation. *Bicovas*, 1, 34-41.

Urabe. V.K. Ikemoito, T and Aida. 1986. Studies on *Sympetrum frequency* (Odonata: Libellulidae) nymphs as natural enemy of mosquito larvae *Anopheles sinsensis* invice fields Z, Evalution of predatory capacity and efficiency in laboratory. *Jap. J. Sanid.* Zool, 37 (3), 213-230.

Wallis, R.C. and Speilman A. 1953. Laboratory rearing of *Culex salinarius. Entomol. Soc. Washington*, 55, 140-142.

Species Index

A

Acrogomphus 38

Ae. (S.) *uniliniatus* 44

Ae. (S.) *vittatus* 43

Ae. dorsalis 47

Ae. fulvus 45

Ae. indica 44

Ae. scapularis 45

Ae. serratus 45

Aedes 19, 27, 32, 35, 42, 45, 78

Aedes (Stegomyia) *aegypti* 35, 43, 50

Aedes (Stegomyia) *albopictus* 35, 43

Aedes aegypti 2, 6, 36, 47, 50, 55, 65, 76, 80, 83, 84

Aedes albopictus 47

Aedes serratus 45

Aedes sp. 29

Amphithemis mariae 40

An. (An.) *barbirostris* 44

An. (An.) *stephensi* 44

An. (C.) *stephensi* 43

An. (C.) *subpictus* 43

An. (C.) *theobaldi* 44

An. cruzii 45

An. mediopunctatus 45

Anax immaculifrons 38, 50, 55

Anax parthenope 38

Ankistrodesmus falcatus 63

Anopheles 27, 35, 42, 45

Anopheles (Cellia) *culicifacies* 35, 43

Anopheles albimanus 72

www.ingramcontent.com/pod-product-compliance
Lightning Source LLC
Chambersburg PA
CBHW070705190326
41458CB00004B/855